ETHNOBOTANY OF THE COOS, LOWER UMPQUA, AND SIUSLAW INDIANS

Ethnobotany of the Coos, Lower Umpqua, and Siuslaw Indians

PATRICIA WHEREAT PHILLIPS

Foreword by Nancy J. Turner

Oregon State University Press Corvallis

Photo on previous page: Twine made of cattail leaf by the author. Photo by Curtis Phillips.

Library of Congress Cataloging-in-Publication Data

Names: Phillips, Patricia Whereat, author.
Title: Ethnobotany of the Coos, Lower Umpqua, and Siuslaw Indians / Patricia Whereat Phillips.
Description: Corvallis : Oregon State University Press, 2016. | Includes bibliographical references.
Identifiers: LCCN 2015049037| ISBN 9780870718526 (original trade pbk. : alk. paper) | ISBN 9780870718533 (ebook)
Subjects: LCSH: Indians of North America—Ethnobotany—Oregon. | Ethnobotany—Oregon. | Medicinal plants—Oregon. | Confederated Tribes of the Coos, Lower Umpqua and Siuslaw Indians of Oregon—Ethnobotany.
Classification: LCC GN476.73 .P55 2016 | DDC 581.6/3409795—dc23
LC record available at http://lccn.loc.gov/2015049037

♾ This paper meets the requirements of ANSI/NISO Z39.48-1992 (Permanence of Paper).

First published in 2016 by Oregon State University Press
Printed in the United States of America

Oregon State University Press
121 The Valley Library
Corvallis OR 97331-4501
541-737-3166 • fax 541-737-3170
www.osupress.oregonstate.edu

Contents

Foreword

This book is a product of love and respect, of dedication and commitment, of recognition and attention to history, people, and environment. Each detail it contains, each plant described, each name in the three Native American languages, is a gift—a crystal or a nugget—lovingly recorded and treasured as a component in a rich system of knowledge. A quotation from the preface, "All things shall grow," surely applies to the book itself. Patricia Whereat Phillips, who is of Milluk Coos heritage, faced many difficulties in its creation, noting, "This project has been like trying to weave a large basket out of small scraps and fragments." Placed in a context of historical loss, ecological degradation, and knowledge erosion, the outcome she achieved is truly remarkable.

The book is a new and expanded version of one published over ten years ago, *Ethnobotany of the Coos, Lower Umpqua, and Siuslaw: Plants Used for Tools, Food, Medicine and Clothing as Remembered by Our Elders and Ancestors*. Resulting from a collaborative ethnobotany project, this original publication was coordinated and compiled by Debra Hall and published by the Confederated Tribes of Coos, Lower Umpqua, and Siuslaw Indians in 2004 (114 pages). Given the ongoing revitalization of cultural traditions such as basketry, berry picking, traditional medicinal plant use, canoe making, and creating and using traditional tools, as well as language revival, this new, expanded volume is so very important and timely. It is even more to be treasured because, to date, there has been such a scarcity of published materials on ethnobotany from western Oregon.

For me, as an ethnobotanist who has collaborated with Indigenous plant experts and practitioners in northwestern North America for over forty years to help document their rich botanical heritage, this book is a celebration of both the distinctiveness of people's knowledge and use of plants, and of the commonalities of this knowledge across time and space. The three language groups included in this book are related; Hanis and Milluk are classified in the Coosan language family, and both are distantly related to Siuslaw, in the Yakonan family. Along with many other languages of the region (Chinookan, Klamath, Kalapuyan, Wintu, Maidu, and the Ts'msyenic languages of British Columbia), these languages are provisionally classed as part of a broader "Penutian" language phylum.

The lifeways of the Coos, Lower Umpqua, and Siuslaw peoples parallel those of many other coastal peoples. Residing in the coastal regions and lower river estuaries, they moved upriver for a period of time each year to fish for salmon and lamprey eels and to harvest a wide range of plants for food, medicine, and materials. The recent histories of these peoples, like those of other Indigenous groups throughout North America and worldwide, have likewise been difficult. Increasingly, as newcomers encroached with settlement and development, the original peoples were confined to smaller and smaller land bases. The Coos, Lower Umpqua, and Siuslaw were encouraged to move inland permanently. People endured cultural losses, food shortages, conflicts, and displacement. With these disruptions, languages disappeared; the last speakers of the Milluk, Hanis, and Siuslaw languages passed away in the 1960s and early 1970s.

Yet, as this book attests, much cultural richness remains, thanks to documentation by linguists and ethnographers like J. P. Harrington, Melville Jacobs, and Philip Drucker, and especially thanks to the knowledge holders and experts of the tribes themselves. The book presents both an open window onto the past and a pathway through which this knowledge can be renewed and revitalized for future generations. The knowledge itself is so compelling. It includes not only the names and habitats and harvesting information for dozens of the region's plant species, but also the methods that people used in the past to promote and enhance the growth and productivity of these species. As described in the following chapters, the techniques for harvesting and managing these plants through digging, tilling, and periodic burning resulted in the transformation of entire ecosystems, the creation of anthropogenic landscapes. As has been the case more widely, the European newcomers were often blind to the role of Indigenous harvesters and managers in creating and maintaining the picturesque open meadows and prairies so sought after by the settlers.

So many of the details recorded for these peoples of coastal Oregon I have learned about from others distantly removed. Somehow these peoples have been both recipients and disseminators of the same ancestral practices that have supported peoples' survival and well-being all along the coast for millennia. The use of fire, for example, to create clearings conducive to camas (*Camassia quamash*) growth and wild berry production is widespread in northwestern North America and beyond. Burning hazel (*Corylus cornuta*) bushes to renew their growth, create long withes for cordage, and enhance nut production is a known practice in British Columbia as well as in Oregon.

Cow parsnip (*Heracleum maximum*) is well known throughout western North America as a springtime vegetable that, once peeled, is sweet and flavorful. Beach lupine (*Lupinus littoralis*) is a known root vegetable of the Haida. Of course, wild strawberries (*Fragaria* spp.) and many of the other berry species are known and enjoyed wherever they grow. Many of the technologies are also more widely used: earth ovens for slow-baking camas bulbs; application of "arrowwood," or ocean spray (*Holodiscus discolor*), for root diggers and arrows; and use of Douglas-fir bark (*Pseudotsuga menziesii*) for fuel and poles for dip net handles, western hemlock bark (*Tsuga heterophylla*) for dyeing fishnets to make them less visible to fish, maple wood (*Acer macrophyllum*) for canoe paddles, and fireweed fiber (*Chamerion angustifolium*) for cordage. Medicinal plants also had common applications, including tall Oregon grape (*Berberis aquifolium*) for blood purification, cascara bark (*Rhamnus purshiana*) as a strong laxative, and bitter cherry bark (*Prunus emarginata*) to treat tuberculosis. Ocean spray, whose blooming is an indicator for sockeye salmon fishing and butter clam harvesting on the British Columbia coast, signifies the time for hunting bull elk in the mountains for the peoples of the Oregon coast.

Spiritual beliefs, such as the teaching that picking flowers can cause rain and bad weather, are likewise widely taught in Indigenous cultures. In the Ditidaht (Nitinaht) language of southwestern Vancouver Island, for example, trilliums (*Trillium ovatum*) are called *chaachaawaʕs* (literally, "sad ones on the ground"), and children are taught that picking these flowers will cause fog. Similarly, on Haida Gwaii, red columbine (*Aquilegia formosa*) is called *dall-sgid* ("red-rain-leaves/medicine") in the Massett dialect, and picking these flowers is said to cause rain, right during a time when people are trying to harvest and dry their edible seaweed. Similarly, the association of tree fungi with echoes extends northward at least to Vancouver Island; the name for tree fungus in Straits Salish is *təw'təw'ə́ləqəp* ("echo" or "echo maker"), and the Ditidaht name, *dayats'uʔ*, also means "echo." The teachings recalled by Daisy Wasson Codding (Upper Coquille/Milluk), to be respectful to fish, animals, or trees, are remarkably similar to those of numerous other First Peoples all across North America.*

Thus, there was, unmistakably, communication among peoples of the Northwest coast and beyond: insights and technologies shared, ideas and concepts communicated, stories and ceremonies adapted to local circumstances. Today, in an ever-changing, ever more globalized world, the idea of continuing to share this rich cultural knowledge among different peoples is appealing. It is a way, in a modern context, of continuing these time-honored traditions, bringing them forward and into the future, where they will be, without any trace of doubt, as important for humanity and for all of the other life forms we share the planet with, as they ever have been. From metaphorical teachings to practical knowledge for sustaining plants and habitats, this treasured wisdom is a gift for the generations to come. Thank you, Patricia, and all those knowledge holders of the Coos, Lower Umpqua, and Siuslaw communities, for bringing it all together.

Nancy J. Turner

* For further information on the plant uses and knowledge of other peoples of northwestern North America, see Turner, Nancy J. (2014).

Preface

Now the two watched their new land.
    The world was without trees.

"How is your heart?
    Shall we stand up some trees?"

"It shall surely be good that way."

Now surely everywhere they stood up eagle's feathers.

It grew in the new land.
    Already high up it grew in the new land.

"Now we shall just watch it."
    Surely the two watched it.

The eagle's feathers, those are the douglas-fir trees.

"All things shall grow."
    (And from there, all other plant life grew).

Adapted from *Arrow Young Men*[1]

The coast of southern Oregon is a botanically rich region; some plants found here are thought of as more part of the Siskiyou or northern California flora, mixed with the flora of the Pacific Northwest. Growing up in Coos Bay, I was unaware of my home's interesting botanical ecology, but I did spend many happy summers wandering the hills above the eastern shore of Coos Bay picking salmonberries, thimbleberries, red huckleberries, and blackberries, as well as raspberries and strawberries from our garden. I loved spending time in the woods. I had grown up being told of my Milluk Coos heritage. I knew that every time I filled a bucket with berries I was connected to a long line of Coos grandmothers who, for generations, had done the same thing. I felt a connection to a place and a family history.

Yet except for berries, I had little interest in plants. When I went to college and studied biology, I focused mainly on animals. All I can recall of botany from those years is frustration in the lab, and being unable to distinguish different plant tissues under a microscope. Animals and their biology were much easier for me to understand.

My interest in ethnobotany began almost fifteen years ago when a college student asked me about traditional medicines. I began to do some research and to my dismay found that there was almost no published information on the ethnobotany of western Oregon peoples. So over the next few years, I talked with a few elders and scoured the ethnological and linguistic notes that had been written on the Coos, Lower Umpqua, and Siuslaw going back to the mid-nineteenth century. The more research I did, the more my interest grew. I compiled a list of dozens of plants used for food, fiber and clothing, tools, and medicine. The Confederated Tribes of Coos, Lower Umpqua, and Siuslaw Indians launched an ethnobotany project, coordinated by Debra Hall (Coos), which culminated in a booklet distributed to tribal members.

The booklet was the first compilation of ethnobotanical information specific to our tribes. But over time, as I found a little more information and gained a better grasp of the tribal languages, I decided I wanted to create a more detailed document. I went back to the original sources and read through them again. I recompiled all the information I could find. I pressed harder to find sources I might have missed before—and I did find some. This project has been like trying to weave a large basket out of small scraps and fragments—often I would find just one small reference to a plant in one document, then another brief reference buried in another anthropologist's notebook. This long and careful search was productive, as I was able to find a few more native words for plants, and with further research I found more information and changed the identification of a couple of plants.

None of the investigators who interviewed native people had any specific interest in ethnobotany. When they do mention plants, the descriptions of individual plants are often vague and contradictory, or the plants are identified by only a common name. I am neither a botanist nor an anthropologist, but after long research I am confident of my identifications of the plants listed here, although there are still a few plants that I have been unable to identify.

With what I have been able to learn through the course of my research, I have gained a great appreciation for my ancestors' knowledge of the natural world, and for the plants themselves. I have come to know their indigenous names, and when I see these plants in the forests and dunes they seem more like old friends than background greenery.

This book lists only a portion of the plants once known and used by the Coos, Lower Umpqua, and Siuslaw. Doubtless there were many other culturally useful plants that were never recorded in the notebooks of ethnographers and linguists, and much of that knowledge was lost to later generations. Nevertheless, what knowledge survives documents an important and too long neglected aspect of our culture and history.

Some cultural practices never died out, particularly the gathering of berries and the making of certain teas. Today there is a renaissance in traditional weaving, canoe making, and tools.

To gain a better understanding of the Coos, Lower Umpqua, and Siuslaw, and to place ethnobotany in a broader cultural context, the following chapters provide a pronunciation guide for the native languages, describe who the informants and ethnographers were, and provide summaries of the culture and history of the tribes.

How to Use This Book

The scientific names and family classifications of species are taken from the Oregon State University Herbarium database, current in April 2014. Common names are also drawn from the herbarium and the northwestern plant guides *Plants of the Pacific Northwest Coast* by Jim Pojar and Andy MacKinnon, *Plants of Western Oregon, Washington and British Columbia* by Eugene Kozloff, and *Handbook of Northwestern Plants* by Helen M. Gilkey and La Rea J. Dennis.

This book is meant to be descriptive of Coos, Lower Umpqua, and Siuslaw ethnobotanical knowledge and practice. It is *not* meant to be used as a how-to book for collecting plants for food or medicine. Some plants that are safe to eat have look-alikes that are quite toxic. Others are safe only when processed correctly. For safety one needs to have detailed knowledge provided by experience and good field-guide books, and to be confident in identifying and processing plants. Knowledge of the land is necessary too—some private and public lands use herbicides and pesticides, and some public land agencies may require permits for gathering certain plants or fungi. In any event, this book is not meant to be a field guide for gathering plants for food or medicine.

Acknowledgments

This book would not have been possible without the inspiration and help of so many people. My dad, Don Whereat, has been a great inspiration in all his work studying Coos, Lower Umpqua, and Siuslaw history and culture. We spent many fun times as a family picking berries, crabbing, and clamming, and he taught me so many things.

Dr. Stephen Dow Beckham has given me lots of support over the years, answering many questions and sending me fascinating notes from Dr. Walton Haydon, a medical doctor and amateur botanist who studied our local flora a century ago.

I also want to thank Reg Pullen, Phyllis Steeves, Sue Townsend, Dan Segotta, Jenny Sperling, Douglas Deur, and Nancy Turner for answering questions and giving me ideas and new leads over the years. I want to thank Debra Hall for her hard work compiling photos and text and interviewing many elders for the ethnobotany booklet by the Coos, Lower Umpqua, and Siuslaw Indians printed in 2004, and Whitney Long, who got that ethnobotany grant started. I also want to thank tribal elders I've talked to over the years—Dorothy (Barrett) Kneaper, Paul Benasco, Ray Willard, Ida Helms, Grace Brainard, Ron Brainard, Dennis Rankin, Lillian Lott, and Sue Olson.

I also want to recognize the Coos, Lower Umpqua, and Siuslaw tribal youth. Over the last few years at summer school you have been enthusiastic and inspiring students (and hilariously brave to try eating crowberries), and you are the ones who rediscovered a wapato patch on a canoe trip on the Umpqua River. Never let go of your sense of wonder, and keep exploring the forests and the waters!

And finally, thank you to my husband, Curtis, and daughter, Morgan, for all your love and support.

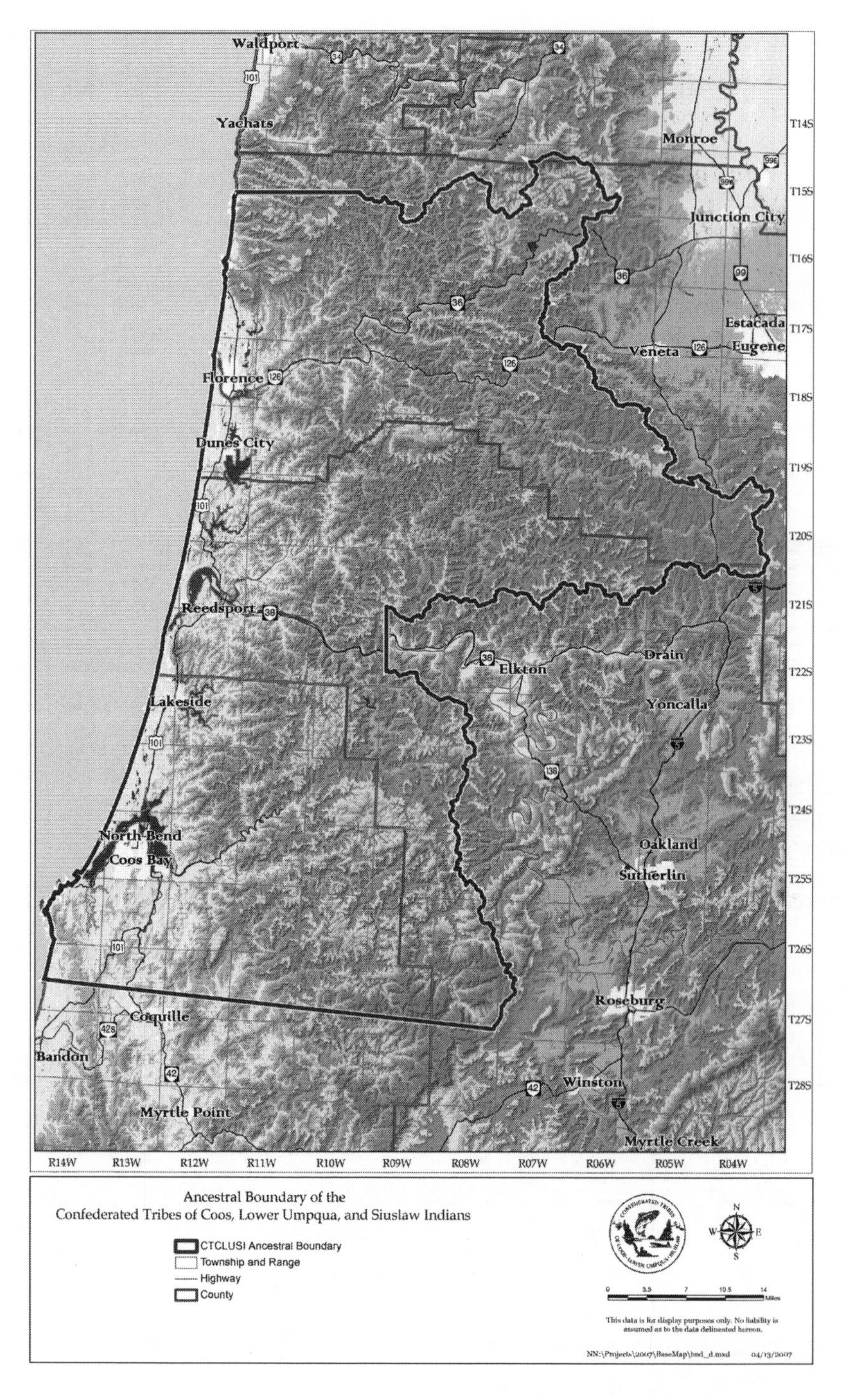

Map of the aboriginal territories of the Coos, Lower Umpqua, and Siuslaw Indians, courtesy of Jeff Stump, Confederated Tribes of Coos, Lower Umpqua, and Siuslaw.

1 Indigenous Languages

Since this book lists the native names of the plants in the indigenous languages of the Coos, Lower Umpqua, and Siuslaw, a discussion of the native languages is necessary. On Coos Bay, two languages were spoken. Milluk was spoken by people living at Cape Arago, South Slough, and lower Coos Bay, and also by the Nasomah (Lower Coquille) people at the mouth of the Coquille River. Hanis was spoken in the Empire district of Coos Bay, and all along the rest of Coos Bay north to Lakeside and Tenmile Creek, which was the traditional border between the Hanis and Lower Umpqua peoples.[1] The Siuslaw and Lower Umpqua spoke mutually intelligible dialects of the same language, usually referred to as Siuslaw (to avoid confusion with other native peoples also known as "Umpqua"). One of the few major differences between the two dialects is that in some words Siuslaw has an *n* while Lower Umpqua has an *l*.[2]

The three languages are related—Hanis and Milluk more closely to each other (traditionally described as the sole members of the Coosan language family), and both rather distantly to Siuslaw (which has been grouped with Alsea/Yaquina in the Yakonan family). All three of these languages, along with Alsea/Yaquina, are tentatively classified as part of the Penutian superstock, which includes many languages of western Oregon and California such as Chinookan, Klamath, Kalapuyan, Wintu, and Maidu.[3]

The neighboring languages on the Oregon coast are completely unrelated. To the south of Coos Bay, dialects of Athabaskan were spoken all along the coast to northern California. To the north of *Tsɪ'ɪmał* (Tenmile Creek in northwestern Lane County) were the Alsea and Yaquina, who spoke a language distantly related to Siuslaw. North of them were the Tillamook, who spoke a Salish language. In the inland valleys east of the Coast Range, Kalapuyan languages were spoken north of the Umpqua divide; Upper Umpqua (an Athabaskan language) was spoken in the Umpqua Valley and Camas Valley.[4]

Multilingualism was common, probably because of the widespread practice of men seeking wives from neighboring tribes and traveling widely for intertribal trade, gambling, and ceremonies. Some entire communities were said to be bilingual—the Lower Coquille people were fluent in both Milluk and southwestern Oregon Athabaskan, and the people of the Winchester Bay village were said to be fluent in Hanis and Siuslaw.[5]

Some entries will have names in only one or two of the native languages, or none at all. Sometimes an informant could not recall the native name when asked, or gave a description of a plant but did not volunteer a native name.

The orthography is one used by many contemporary linguists and anthropologists. To assist the reader who is unfamiliar with it, the following pronunciation chart is provided.

	Consonants
c	Like *ts* as in *cats*. Unlike in English, the *ts* can occur at the beginning of a word.
č	Like *ch* as in *church*.
tł	Like *tl*. It has no equivalent in English. Its voiced counterpart, dl, is in English words like *puddle*.
q	Like a *k*, only pronounced farther back in the mouth.
’	This is the glottal stop. It is a break in the flow of air made by closing one’s glottis. Say *uh-oh*—the break between the syllables is a glottal stop. When it appears next to a consonant—c’, tł’, q’, p’, č’, t’, and so forth—the consonant is pronounced with an accompanying “pop.”
š	*Sh* as in shell.
ł	Essentially a voiceless *l*, like Welsh *ll*. To make this sound, shape your mouth just as you would to make an English *l*, but instead of voicing it, blow air.
x	A raspy *h* sound, similar to the *ch* in German *Bach*.
x̣	A raspy *h* sound pronounced farther back in the mouth.
ɣ	The voiced equivalent of *x*.
	Vowels
a	As in *father*.
i	Long *ee* sound as in *bee*.
u	Depending on the surrounding consonants and vowels, this can be pronounced like the vowel sound in *boat* or *boot*.
æ	As in *cat*.
ə	Neutral vowel like the *a* in *about*.
ɪ	Like *bit, hit*.
ʊ	Like *book, hook*.

Silver Falls, Golden and Silver Falls State Park.

2 Cultural Background and History

That's the only way they've been talking.
        They didn't come from any place.
That was their only place.
        They didn't know where they came from.

Every stream has people on it.
        That's how they all had a stream.
That's the way they know themselves.
        All other tribes had their stream as their land.
                —Jim Buchanan, Hanis Coos[1]

The Coos, Lower Umpqua, and Siuslaw peoples shared many cultural traits. These tribes have a long history of trading and intermarrying. The principal villages were concentrated on the estuaries and coast, but there were also small villages upriver above tidewater. People from the coast and lower estuaries moved upriver seasonally to fish for salmon and lamprey, as well as to gather plants for medicine, food, and basketry. Each village was politically autonomous, with its own headman, or chief. The village headman was the wealthiest man in the village. Wealth was counted in valuables such as strings of dentalium shells, sea otter hides, scalps of red-headed woodpeckers, and other rare items. There were four social classes: the wealthy, the common people, the poor, and slaves. Slaves were usually captured in battle or traded from other tribes. People had to be on the lookout for slave raiders from Tillamook and the Columbia River.[2]

There were several types of houses. The houses of the wealthiest people were semisubterranean, with red cedar planks used to build the roofs and line the walls. A more modest house might have only dirt walls and a red cedar plank roof. Temporary camp structures and the houses of some poor people were wholly aboveground, built with thatched roofs and walls of sword fern and bulrush. People also built storage sheds to house their tools, men's and women's sweat lodges, and brush fences to create windbreaks to make a comfortable outdoor work space.[3]

Canoes were important and were taken everywhere, on offshore fishing expeditions and far upriver. They were even portaged some distance. The Coos Bay people used the chain of lakes through the sand dunes to portage canoes from the bay to Tenmile Creek.[4] The Siuslaw and Lower Umpqua poled canoes far up creeks on hunting and trapping expeditions. They carved canoes mainly from red cedar logs, but they occasionally used Port Orford cedar, washed-up redwood logs, and even spruce.[5]

The people had access to a rich array of resources, from the sea to the Coast Range. They learned from an early age to observe their environment and note when and where certain plants and animals could be found. The near shore and estuary teemed with a variety of fishes that were caught with nets, baited floats, weirs, and trap baskets. Women were the principal gatherers of shellfish and other tide pool delicacies. The annual runs of salmon and lamprey (colloquially known as eels) were followed from the estuaries to upriver fishing camps. The men hunted ducks, seals, sea lions, deer, and elk. These animals provided not only food and hides but also bones and antlers, which were the raw materials used to create tools—knives, fishhooks, harpoon points, and adzes. Whales were usually obtained only when they were trapped in a slough or washed up on a beach.

The botanically diverse Oregon coast provided a wealth of edible greens, berries, nuts, and roots. The gathering of plant foods was done primarily by women, although sometimes men assisted on large root-gathering expeditions. Berry gathering in the Coast Range could involve large numbers of people. Methodist missionary Gustavus Hines estimated on his arrival to the lower Umpqua River (near Winchester Bay) that there were three villages on the lower river, and that "the whole number, as near as we could ascertain, amounted to about two hundred men, women and children, about one-third of whom were absent in the mountains, for the purpose of gathering berries."[6] The art of basketry was also primarily the province of women, although men would make fish trap baskets and nets.

Men made the hunting equipment—arrows, bows, fishing spears, and the like. Men were also the woodworkers who made the canoes, paddles, planks, and house posts.

There were medicines known by all people, many but not all of them derived from plants. Red ochre mixed with fat was used to heal cuts and pimples, and women painted their faces with it to protect their skin from the sun and wind when out gathering clams and mussels at the shore.[7] Powdered sand dollar or snail shells were used to heal newborns' belly buttons after the umbilical cord fell off, as well as other sores.[8] Burns were treated with an application of a nursing mother's milk. Sometimes aching joints were treated by making shallow cuts in the skin and rubbing the blood on a stick, which was then thrown into a river.[9]

Water itself was considered healing. Aching joints and wounds were bathed in cold freshwater, or sometimes pools of seawater in rock depressions. There were medicinal springs, called *šqálya* in Hanis, with water that was given to sick people to drink.[10] Drinking seal oil kept away the chill in winter and also treated sickness in the stomach and lungs. Most healing came through the medium of shamans, or doctors. Doctors were believed to pull illnesses out of their patients. The sickness would manifest as a worm, snail, or bit of slime that the doctor would destroy by drowning it in water or eating it. Doctors' spirit powers sometimes gave them knowledge of plant medicines, and people would obtain tonics from them.[11]

In the indigenous view, everything has a spirit. Spirit powers were sought by both men and women to provide good fortune and guidance. Most people did this seeking around adolescence, hoping to find a power that would be with them for the rest of their lives. The spirits revealed themselves in dreams. Most people obtained only one or two spirit powers. If a person had a good dream, one he or she wanted to keep, that dream was kept private. People never revealed the identity of their power or powers. Only when a person died would the family learn what the power was. If a person had a dog power, dogs barked when he died; winds blew when someone with rainbow power died, and so on.[12] If a person revealed what his or her power was, the power would depart (although this changed with the introduction of the Warm House dance to the Oregon coast in the 1870s, because revealing dreams was a central part of this religious movement).[13] Those who trained hard and acquired several powers could become doctors. Doctors performed cures by removing illnesses from their patients, but this power was also regarded as dangerous. People believed that doctors caused most illnesses in the first place. For a fee, some doctors could be hired to cause someone to become sick. A doctor's powers also revealed certain medicines, and people could go to doctors to obtain tonics and other treatments.[14]

Another kind of spirit power, the luck powers (known as *tɫxínxat* in Milluk, *tɫxínəx* in Hanis), brought good luck and wealth. Certain individual plants were also luck powers. Annie Peterson recounted advice she had learned from elders about wealth powers:

> This is the way we always speak when we tell our children about it. No matter how bad it (the encounter) be, even if they are not good encounters. Even then they tell their children that way.
>
> "Go round about outside! Fear nothing! No matter how bad (fearful) it may be, you are to go nevertheless right to it here, (perhaps) to the ocean, (or perhaps) to a lake, no matter how bad it may be. You must not fear it."
>
> "If there are tree snags (or stumps) at the lake, if there is a black huckleberry bush on it, you should swim to it there. It indeed will make a fine hand game stick. You might become rich with it, if you encounter such a thing."
>
> Even though (they are) young girls they will nevertheless tell such things to them. And indeed that is what they (girls) themselves do. That is the way a girl at puberty goes around, swims, and encounters a luck power person indeed.
>
> She might obtain money (with her encounter power) she might obtain a husband with it. That is what makes them

> become wealthy. That is the way the children believe their parents.[15]

If a person was lucky enough to encounter a luck power in the form of a bush or tree, its wood would make a very lucky implement in gambling games or sports. Jim Buchanan mentions another plant, "ironwood" (ocean spray, *Holodiscus discolor*), that could also appear as a *tłxínxat*:

> Ironwood was a power, a woman, weeping out in the woods, all decorated. He saw it was a person, he went, 4 times he didn't get to her, the 5th time, scared to death, he gets to the weeping woman. She had a white thing which he took from her nose. Then he fainted, lost consciousness. When he came to she was gone. Neither had spoken. While lying senseless he dreamed. In the dream she told him he would be wealthy. While he lay there it became 2 dentalia. He brought it home, he cached it 5 days in a hole in moss in the ground, and it multiplied. She was ironwood.[16]

The wood of ocean spray was often used to make game sticks and tallies for the hand game, which was played throughout western North America.

The first contacts with nonnatives that the Coos, Lower Umpqua, and Siuslaw had were brief ones with passing ships in the late eighteenth century. In the early nineteenth century, fur trappers made irregular treks through southern Oregon. In 1836, the Hudson's Bay Company built Fort Umpqua at what became the town of Elkton.[17] Jean-Baptiste Gagnier ran the fort for most of its existence, and he was married to a Lower Umpqua woman, probably to help maintain friendly relations with the tribe. The fort did become a regional magnet for trade. Lottie Evanoff recalled that her father traveled there to trade furs for manufactured clothing. "When my father was young he walk[ed] to Allegany & from there over the hill to Scottsburg, packing beaver hides, etc. He got pay. My father said: Those were the first clothes I ever put on, I got a belt. . . . Old man Ganyar [*sic*; Gagnier] was who was buying all kinds of furs when my f[ather] was a little boy."[18]

In the 1850s, settlers began moving to the southern Oregon coast. The government negotiated a treaty with the coastal tribes in 1855, but it was never ratified. When war broke out with the Rogue River people and their allies, the Coos Bay Indians were rounded up and moved up to a military fort hastily built on the north spit of the Umpqua River. The Coast Reservation was created by a presidential executive order, and many of the Coos and Lower Umpqua Indians were moved to the Alsea subagency at Yachats. Indian women who were married to white men were often (but not always) permitted to remain off the reservation. Many of these families sheltered other Indians who ran away from the reservation, or helped hide them from soldiers who went out on sweeps to capture runaways.[19]

The Siuslaw people stayed on the lower Siuslaw River, which was initially regarded as part of the reservation. This arrangement was short-lived. Over the years, the Coast Reservation was whittled down in size, and the native people progressively lost more land. In 1877 the Alsea subagency was opened to settlement and the Indians were encouraged to move inland to the Siletz Agency.

The native people began farming introduced crops fairly early. In 1858 Indian agent E. P. Drew wrote to the Indian Affairs office in Washington that the newly proposed Coast Reservation should be extended south from Siltcoos Creek to the Umpqua and Smith Rivers to include the Indians' new farm plots:

> I suggest, that the Southern boundary be changed to the Umpqua and Smith river, in which event, a country well adapted to the wants of a few Indians will be acquired by them, and the Military Reserve will join the Southern boundary of the Indian Reservation. Agricultural lands, between the Siuslaw and Umpqua rivers, will thus be added to the Reserve, which, when properly cultivated, will be sufficient to subsist them. . . . They incline in this mode of farming, and many of them have now small parcels of land, which they are cultivating to a limited extent on Smith's rivers [*sic*], and about the lakes between the Siuslaw and Umpqua rivers.[20]

When later removed from Fort Umpqua to Yachats in the fall of 1859, the Coos and Lower Umpqua again planted gardens, but not always with success. In October 1860 a report to the Bureau of Indian Affairs stated that the crops had failed entirely.[21] There would be crop failures in other years. The Indians fished the Yachats River and other creeks on the reservation and gathered mussels, but at times they went hungry. Frank Drew, a Coos man raised at Yachats, recalled,

> At the Yahats [*sic*] Reservation they had suffered hardships for want of food. They lived mostly on mussels and fish. Many of them were drowned . . . while trying to catch fish or gather mussels, washed away by the breakers. During 1861–1876 the Coos were so cruelly used by the Indian agents one of those name was George Collins. The Indians that ran away from the Reservation to their former homes were brought back by force—young or old—and flogged and whipped unmercifully.[22]

Annie Miner Peterson also recalled hard times at Yachats:

> Now we returned to Yáhatc, and we lived there with my mother's own relatives. We never went to Yaquina again, I suppose I was about six years old when we left Yaquina. We stayed at Yáhatc. We lived poorly, we had nothing, we had no food, only just some Indian foods. That is how we lived at Yáhatc. . . . We had no clothes, we had to wear any old thing. That is how I grew up.[23]

At Yachats the native people raised potatoes, carrots, rutabagas, onions, artichokes, and peas. The agency buildings included a barn and a potato house to store crops.[24] Potatoes were not just food; the starchy water they had been boiled in was used to wash feet in winter to protect the skin from frost.[25] The Indian agent had his own personal garden—once Frank Drew and some other boys raided his strawberry patch. Frank was caught and as punishment had to spend a day locked up in the potato house.[26]

They also tried to raise wheat and oats at Yachats. The climate of the Oregon coast is not particularly amenable to wheat farming, but nevertheless for many years they planted it. Some of the Indian agents recognized the futility of this exercise. In 1873 agent T. B. Odeneal wrote, "The sub-agency is on the sea shore, on a bleak plain. . . . The incessant gales render the climate very disagreeable, and at the same time serve to keep the vegetables saturated with salt water spray, or mist, which is very damaging to the grain crops, always blighting and sometimes killing the crops entirely."[27]

The Siuslaw people were still living along the banks of the Siuslaw River, where they had their own fisheries and gardens. In 1862 Indian agent Linus Brooks noted that they raised "comfortable supplies of potatoes, corn, squashes, carrots and peas."[28]

From 1860 to 1877, over half of the Indians at Yachats died.[29] When the agency was closed in 1877, many of the Coos and Lower Umpqua people refused to go to the Siletz Agency and went south to live on the Siuslaw River, while a few returned to the Umpqua, and many went to live at Coos Bay, mostly around South Slough.

Many intermarried with non-Indians and found work in canneries, on fishing boats, and in logging.[30] They never forgot the unratified treaty of 1855 and as early as 1890 began pushing for a land claims settlement.[31] Since that time, the Coos, Lower Umpqua, and Siuslaw peoples have worked together on issues of land payments, termination, and restoration. After decades of meetings, the tribes received permission from Congress to sue the federal government for land claims. Several elderly Indians and pioneers gave their depositions in 1931.[32] The court did not release a decision until 1938. The court ruled against the Coos, Lower Umpqua, and Siuslaw, stating that an unratified treaty and testimony by people who had an interest in the outcome of the case were insufficient proof of aboriginal possession. The tribes tried to appeal to the US Supreme Court but to no avail.[33]

In 1937, the Bureau of Indian Affairs commissioned a meeting hall to be built on six acres in Empire (now part of the city of Coos Bay).[34] This hall became a center for tribal gatherings, meetings, and weekly medical care.

In the 1950s, the federal government began to aggressively pursue the policy of termination. By termination the government wanted to withdraw the recognition of tribal governments, dissolve the collective tribal ownership of reservations, and make Indians full citizens of the

United States while eliminating their own tribal governments.[35] The Coos, Lower Umpqua, and Siuslaw vehemently opposed this, but to no avail—Congress passed a bill (Public Law 588) on August 13, 1954, terminating all the tribes and bands of western Oregon. The law took effect two years later.[36]

Even in the midst of the termination process, the Coos, Lower Umpqua, and Siuslaw fought it and continued to press for a land claims settlement. Then-chairman Howard Barrett Jr. lobbied Senator Wayne Morse on both of these issues,[37] but the tribes had no success on either front.

In spite of all these setbacks, the tribes regrouped and pushed for re-recognition by the federal government. On October 17, 1984, President Ronald Reagan signed the restoration bill recognizing the Confederated Tribes of Coos, Lower Umpqua, and Siuslaw.[38] This was part of a reversal of 1950s federal Indian policy. In a twelve-year span, from 1977 to 1989, other Oregon tribes that had been terminated in 1954 were also restored: Confederated Tribes of Siletz (1977), Cow Creek Band of Umpqua (1982), Confederated Tribes of Grand Ronde (1983), Klamath Indian Tribes (1986), and the Coquille Indian Tribe (1989).

3 The Ethnographers and Their Informants

> Two rivers (Siuslaw and Umpqua) will have one language. Thus the world will be started. One woman and one man I shall send (at a time). . . . (The People living on) two (different) rivers will understand each other's language. Ye will multiply there. (Living on) two (distinct) rivers, (ye will) understand each other's language. . . . Thus the tribes were created.[1]

There were few ethnographers who worked with the Coos, Lower Umpqua, and Siuslaw peoples in the nineteenth and twentieth centuries, and they worked with even fewer informants. Not many of these ethnographers took much of an interest in ethnobotany. One notable exception was James Owen Dorsey, who worked at the Siletz Reservation in 1884. Dorsey worked with speakers of southwestern Oregon Athabaskan, Siuslaw, Alsea, and the Lower Coquille dialect of Milluk. He collected samples of about fifty plants and got names for them in three dialects of southern Oregon Athabaskan and Alsea, but he seemed to be interested mostly in obtaining the native names for the plants. Curiously, he wrote very little about their cultural uses.[2]

Much of the information comes from four Coos Bay informants, Jim Buchanan, Frank Drew, Annie Miner Peterson, and Lottie Evanoff; two Lower Umpqua informants, Louisa Smith and her son, Spencer Scott; and two Siuslaw brothers, Clay and Henry Barrett.

James Buchanan was a Hanis Indian, born at Coos Bay in the village of *Wu'álach* sometime between 1845 and 1848. In 1856, he and all Coos Bay Indians were moved to *Kiwœ́'œt*, a point just below Empire, in preparation for being moved to Fort Umpqua, a military fort established on the north spit of the Umpqua River.[3] The soldiers and Indian agents issued English names to all the Indians. Some unknown agent or soldier, in an apparent moment of humor, named a young boy James Buchanan, after the man elected president that year. It was a name he kept for the rest of his long life. His Hanis nickname was *Cǽtœł*, meaning "worn-out knife," or "knife worn small."[4] In 1860, he was removed with the rest of the Coos and Lower Umpqua to Yachats, which was then part of the Coast Reservation. He became a leader among the Coos Bay people. After the Alsea subagency at Yachats was closed, he moved to the Siuslaw River. He married Eliza, sister of a Coos chief named Jack Rogers.[5] The Coos, Lower Umpqua, and Siuslaw peoples began working on land claims as early as 1890, which culminated in a land claims trial in 1931. Jim Buchanan testified at the trial in Hanis with the help of translators. During his lifetime he was also a noted storyteller and worked with three different linguists to record the Hanis language.[6] For many years he lived at Siboco, on the south shore of the Siuslaw River. Siuslaw elder Dorothy Kneaper recalled that when she was a child, her family would canoe over to Jim Buchanan's house. They brought him salmon and in return he gave them apples from his orchard.

The first linguist he worked with was Henry Hull St. Clair, a student of Dr. Franz Boas. In 1903 St. Clair worked with Jim Buchanan, Tom Hollis, and George Barney to record some vocabulary in Hanis and Milluk. He worked mainly with Buchanan, ultimately recording thirteen myths. Six years later, Leo Frachtenberg came to the Siletz Reservation and worked with Jim Buchanan, Tom Hollis, and Frank Drew. Frank Drew was born to a Hanis Coos woman at the Alsea subagency but had spent most of his life living on the Siuslaw River. He was fluent in English and Hanis and knew quite a bit of the Siuslaw-Umpqua language as well. Frank Drew was quite willing to work with Leo Frachtenberg, but he was very poor at providing texts, which Jim Buchanan could do. So Frachtenberg worked obtaining stories from Buchanan, with the help of Frank Drew as translator.[7]

During this same period Leo Frachtenberg also worked on the Siuslaw-Umpqua and Alsea languages. In 1911 he worked with a Lower Umpqua woman named Louisa Smith (her father, Sunk-in-the-Water, was a Lower Umpqua from the village of *C'álila*, her mother from an upper Siuslaw village), and her husband, William Smith. Her Siuslaw-Umpqua nickname was *Tłma'qt čítł*, Short-Hand, because she was a short-statured person.[8] William Smith was Alsea, but he was also fluent in the Siuslaw-Umpqua language. Louisa Smith had been J. O. Dorsey's informant on the Siuslaw language in 1884. Frachtenberg obtained several texts, but he was not satisfied with their quality, as in his opinion neither of the Smiths were good storytellers, and Louisa was unwell and her mind wandered.[9]

In 1932 Melville Jacobs, a linguist from the University of Washington, began work with Frank Drew, and to a lesser extent Jim Buchanan, to record ethnographic and linguistic information. He also recorded several songs from both of them on wax cylinders. Jim Buchanan died shortly after his work with Jacobs, who was not satisfied with Frank Drew as an informant. Jacobs occasionally wrote negative comments in the margins of his notebooks from his work with Drew. In Jacobs's opinion, Drew was too acculturated. Drew was also not well liked or trusted by some of his neighbors. People gossiped that he was greedy and scheming and tried to cheat other Indians out of their allotments.[10]

The following two summers, Jacobs worked with Annie Miner Peterson. She was born in 1860 to *Mə́təlt* (her English name was Matilda), of Hanis and Lower Coquille ancestry, and a white father named Miner. Matilda was from one of the Hanis villages in Empire, *Ntisœ'ɪch*. Mr. Miner worked in a sawmill and was rarely home. When Matilda's family was all removed to Yachats, she took her infant daughter and left for Yachats. Annie's first language was Hanis, but she learned Milluk at a young age, and as a young adult she learned English. Jacobs was delighted to find someone who was fluent in both Hanis and Milluk, and he obtained many texts from her in both languages. In addition to texts, he also obtained a lot of ethnographic information. Annie Miner Peterson was a skilled basket weaver, and she gave Jacobs a lot of information about the different types of baskets, designs, materials, and dyes. However, like Frank Drew, she was a controversial figure among some of her peers, and one of her nicknames was *tsmixwn*, meaning "tricky" or "trickster." She dictated her autobiography to Melville Jacobs in the Milluk language, and she is also the subject of a biography.[11]

In 1933 Philip Drucker, then a graduate student at the University of California at Berkeley, visited the Siletz Reservation to do a cultural survey. He worked primarily with speakers of southwestern Oregon Athabaskan languages, but he also worked with Tillamook, Alsea,

Molalla, Lower Umpqua, and Coos Bay Indians. He eventually published brief ethnographic sketches of the southwestern Athabaskans and Alsea. He never published any of the notes he collected from his interviews with the Coos Bay informants, Frank Drew, Annie Miner Peterson, and a Hanis Coos woman named Agnes Johnson. Annie Miner Peterson and Agnes Johnson provided Drucker with the names of several edible plants, but unfortunately Drucker did not record much of a description of the plants, and so some are unidentifiable.[12]

In 1942, John Peabody Harrington worked with some speakers of Hanis, Milluk, and Siuslaw. For Hanis, he worked primarily with Frank Drew and Annie Miner Peterson's niece Lottie Evanoff. Lottie was the daughter of Annie's older sister, Fanny, and Chief Doloos Jackson. Harrington also got a few Hanis words from Martha Harney Johnson. Martha was a full-blooded Hanis Indian who grew up on the Siuslaw River. He got very little Milluk—a few words that Lottie Evanoff remembered, and a few words from Milluk/Upper Coquille sisters Daisy Wasson Codding and Lolly Wasson. For the Siuslaw/Lower Umpqua language, he got many Siuslaw terms from Frank Drew and Spencer Scott, who was the son of Frachtenberg's Lower Umpqua informant Louisa Smith. He also got some Siuslaw vocabulary from brothers Clay, Howard, and Logan Barrett. Harrington had a wonderful ear for language, and his notes are very detailed phonetically. He published none of his work from western Oregon, but his field notes are available on microfilm from the National Anthropological Archives in Washington, DC. He was the last investigator to obtain a lot of detailed information on the Coos Bay and Siuslaw languages and culture.[13]

In the 1950s and 1960s, a few sound recordings and word lists were made, but very little ethnography was captured. During this time, the last elderly speakers of Milluk, Hanis, and Siuslaw passed away. The last survivor was Martha Harney Johnson, who died in 1972.[14]

4 Plants and the Traditional Culture

pa'naq, garden. . . . The Indians here used to have gardens. When [he] first came to his senses he saw all the Indian families having good gardens and also put in a lot of berries. It seemed that the land was fresh in those times, and things now need fertilizer.

—Spencer Scott[1]

The Coos, Lower Umpqua, and Siuslaw territory is in the southern Pacific Northwest region. It stretches in the north from Tenmile Creek near the border between Lincoln and Lane Counties, south to Whiskey Run Creek just to the north of the Coquille River, and east to the summit of the Coast Range. The coastline has rocky promontories and headlands at Cape Arago in the south and Heceta Head in the north. Between these headlands is a sheet of dunes fifty-five miles long. The three major rivers that flow to the coast in this area are the Siuslaw, Umpqua, and Coos. Only the Umpqua has its headwaters in the Cascades and punches through the Coast Range; the other two are comparatively short rivers with their headwaters in the Coast Range.

The coastal region receives an average rainfall of sixty-five inches per year, although there are sites in the Coast Range that receive up to one hundred inches per year. Most of the precipitation falls in the winter and early spring months, but summer cloud cover and fog help keep the region damper than the relative lack of summer precipitation would otherwise indicate.[2] This precipitation supports a rich temperate rain forest flora.

Knowledge of individual species of plants gathered for food, medicine, fiber, or tools, as well as knowledge of land management practices that would enhance the productivity of these important plants, was important in the traditional culture of the Coos, Lower Umpqua, and Siuslaw. Many people are under the mistaken impression that indigenous people did not practice land management, that they simply wandered about the forests and picked whatever they found. Native people intentionally managed the landscape to maximize productivity of desired plants, primarily by lighting well-timed fires and aerating the soil annually by digging bulbs from meadows. About digging sticks and collecting bulbs, one ethnobotanist observed that "their power to transform landscapes had generally been underestimated. The digging stick, for example, in the hands of thousands of women, could turn over and aerate large areas of soil in meadows, coastal prairies, or valley grasslands—greatly affecting the composition and densities of the species found there."[3] Many edible "roots" consist of bulbs and corms, and during harvest bulblets and small corms get dispersed, helping the plants reproduce. Tobacco seeds (*Nicotiana quadrivalvis*) were scattered in well-tended plots surrounded by brush fences.[4]

Through the course of a year, people watched the seasonal cycles and planned their activities accordingly. The names of the months were forgotten, but some recalled that people kept track of the northward and southward movement of the sun. They also used stick bundles to keep track of days, by subtracting one stick from a bundle each day.[5]

Winter was the time to eat lots of stored foods. The storage baskets and rafters were full of dried meat, sea lion stomachs containing seal oil, and nuts, seeds, bulbs, camas cakes, and dried berries. Many of the dried foods were mixed with seal oil to make them richer and more palatable. There were few fresh foods available in this season (steelhead, flounder, and, for those who did not mind braving the cold and rain, shellfish). People spent this season dancing in their houses, eating their preserved foods, listening to stories, and gambling on the guessing game and other "dice" games.

Spring was an especially joyous time of year, because after a long winter of eating mostly preserved foods, at last fresh greens became available and the flowers promised a harvest of fresh berries and nuts to come. Runs of eel (Pacific lamprey) lasted from May to autumn. Summer brought a crop of berries and camas, as well as runs of smelt, jack salmon, and some offshore Chinook salmon. Late summer and autumn brought elk and deer hunting season, fall salmon runs, more berries, nuts, fern rhizomes, and wapato.

The question of when, where, and for how long people moved seasonally is unclear. Frank Drew recalled that people stayed a month or two at summer fishing camps and incidentally hunted elk and deer at that time also. "One's whole family would go along on these excursions."[6]

Annie Miner Peterson said that some Coos Bay people also had autumn fish camps, set up near rapids, falls, and shallow water, where people stayed until midwinter before returning to villages near the coast. "The salmon were when they first came in caught from the lower bay permanent houses. Then as the salmon went up the people went up to their upriver houses, and stayed there for eels and salmon until New Years or thereabouts."[7] However, these upriver villages were never entirely abandoned. A few people preferred the quieter, sheltered existence of these upriver retreats and lived there year-round. Also, according to Annie, not all Coos villages had upriver encampments. The people of the sloughs were able to catch fish and hunt elk and deer near their homes. Only people on the main bay, and the Milluk people of Cape Arago, regularly trekked to upriver camps.

The seasonal rounds of the Lower Umpqua and Siuslaw are even less well understood. The Siuslaw people made camps at riffles up Lake and Indian Creeks, and some traveled to Triangle Lake to socialize and trade with Kalapuyans.[8]

One important fishing site for the Lower Umpqua people was Smith River Falls. In summer and fall, people fished for eels and salmon. Then in autumn they headed back downriver, stopping at a Rain Rock with prayers to bring the winter rains. This was also about the time that fires were set to clear brush and invigorate the growth of other plants.[9]

The location and timing of fires were carefully chosen. Every autumn the camas meadows were fired to clear them of competing brush and fertilize the soil for future growth. Every few years hazel patches were burned. One hill on the south side of Coos River was noted for its productive hazel patch, which was burned off in August approximately every five years.[10] This reinvigorated the growth of the shrubs for use as food and encouraged the growth of long, straight switches for basketry. It also kept down the population of pests that might otherwise have attacked and weakened the wood. The firing of beargrass (*Xerophyllum tenax*) was a nearly universal practice, because the leaves that grew back were a stronger, tougher material for weaving.[11]

Many Pacific Northwest tribes did the same for berry patches, and this was also probably true along the southern Oregon coast. People of the inland valleys fired tarweed (*Madia* sp.) meadows in late summer to burn off the sticky seed coatings. Women then gathered the seeds into baskets with seed beaters. Tarweed seeds were an important food for native people of the inland valleys, whose English name for them was "Indian oats," and the seeds were traded with coastal peoples, where tarweed was uncommon or absent.[12]

Fires were also set to keep underbrush down to make deer and elk hunting easier.[13] These fires were generally of low temperature compared to our highly destructive contemporary forest fires. The annual fires burned out the forbs and brush, but mature trees were unharmed. Indeed, these fires were beneficial in that they acted as a control on insect populations. Oak groves that were regularly burned had fewer pests feeding on acorns, thus improving the food source for native people. Hazel nuts probably benefited similarly.

Plants with edible roots abounded and as women dug them up, they aerated the soil and scattered bulblets and seeds, reinvigorating the growth of several culturally valuable plants. Unfortunately, little has come down to us about indigenous horticultural practices in western Oregon. Such practices are better known and studied among some British Columbia tribes, where natives were aware of the positive impacts of digging to aerate soil, clearing ground with fire, and using fire ash or, in some cases, seaweed, for fertilizer. In California, many tribes also practiced selective burning and weeding, and they scattered seeds and bulbs when digging meadow food plants. Coos, Lower Umpqua, and Siuslaw people probably followed similar practices. It is known that some coastal people fertilized their tobacco (*Nicotiana quadrivalvis*) plots with sand and salmon bones.[14]

Even though Coos, Lower Umpqua, and Siuslaw informants recalled few precontact horticultural practices, there are some tantalizing hints in the language. Words for "garden" appear in Siuslaw (*pa'naq*) and Hanis (*tłxáni*).[15] These words do not appear to be borrowed from either English or Chinook jargon, and they probably referred to tobacco plots, which were cultivated in fertilized fenced-off areas.

Another interesting hint that the native people applied the concept of "garden" to plants other than tobacco comes from a speech made by Siuslaw Dick at a meeting with Indian agents on June 17, 1875. The purpose of the meeting was to convince the residents of the Alsea subagency to move to the Siletz Reservation, a policy universally opposed by the native people who spoke that day. Siuslaw Dick said, "Long time this has been my country. My heart has been heavy on that account. It was never my wish to give up any country. . . . I will not go to another country to die. *Before I ever saw a white man I raised produce on my land.*"[16] It is possible he was speaking of raising potatoes and other crops introduced through trade, or he may have been referring to more ancient ways of caring for camas, springbank clover (*Trifolium wormskioldii*), and other indigenous food plants.

As mentioned in chapter 2, as early as 1858 the Coos and Lower Umpqua were raising potatoes along the Umpqua and Smith Rivers, and within a few years they were raising several other introduced crops. Indian agents noted that the Coos, Lower Umpqua, and Siuslaw were working diligently at farming. Madonna Moss believes that the Tlingits' experience of raising tobacco gave them the skills to successfully raise potatoes.[17] If so, this was probably also true in western Oregon.

But plants were not just objects to be used. It was believed that all living things (and some nonliving things, such as certain rocks and mountains) had a spirit. In Hanis Coos, plants were said to have *tłədœwəs*, life. All things had to be treated with respect, or bad luck might follow. Annie Miner Peterson recalled that she was often scolded as a child for picking up shells and flowers. "You weren't allowed to pick flowers, or pick shells, or take up rocks. They said it might start rain or bring bad luck; things were put there to be there, not to be picked."[18] Daisy Wasson Codding (Upper Coquille/Milluk) recalled that her mother taught her to be respectful to the natural world. "Children were not allowed to make fun or speak disrespectfully of fish or animals or trees."[19]

Trees, more so than other plants, were imbued with spiritual and mythological significance. There were stories of doctors (shamans) who dreamed of a tree appearing as a red-eyed person who brought knowledge of medicines. And in some stories, people gained a tree spirit as their dream power. Sometimes doctors said they received from their spirit powers knowledge of plant medicines, tonics, and other treatments for certain ailments.

Not only trees had spirits; certain individual stumps were imbued with power as well. There was a trail from the Millicoma River to Scottsburg (near the site of the Lower Umpqua village of *C'álila*), and at one point along this trail there was a great old hollow stump. People left offerings such as clothes or beads whenever they passed by. There was a story that once when a party camped near the stump, a young man loudly declared that it was foolish to give anything to a stump. He grabbed a tree limb and hit the stump several times. The next morning, his traveling companions found the young man dead. It looked as though he had been beaten.[20]

Such stumps were not unique to the Coos Bay people. When the Coos and Lower Umpqua were exiled to the Alsea subagency, they learned a story from the Alsea people about a special stump north of Yachats. When people passed by this stump they jumped around it five times and left something like a handkerchief or other small offering inside it.[21]

In a land rich in trees, men had much to choose from for carving household and hunting implements. They split logs with wedges made of stone, hardwood, or antler. They made chisels and adzes from polished stone. The southern Oregon coast did not have a carving tradition as elaborate as that in British Columbia and southern Alaska. The tradition has nearly become lost, but at one time people did carve club handles and other objects. In the 1920s, at Graveyard Point near the mouth of the Coos River, fishermen's nets snagged on the remains of a substantial weir. Some of the weir poles were three to four inches in diameter and ten to eighteen feet long, and they had carvings of men and fish on them. Unfortunately, such elaborate weir poles have not been found since those were removed.[22]

Wooden bowls, troughs, and canoes were first roughed out with fire and the burned material was scraped away. The wood was then finished with carving, polishing, and smoothing with sandstone and finally horsetail (a siliceous plant). Wooden bowls and troughs were often made from burls from any available tree. Some bowls were used as chamber pots during the night, and some troughs were used to bathe small children. Older children and adults generally preferred to bathe in creeks. Canoes were carved mainly from the two locally available species of cedar, western red and Port Orford, and occasionally from spruce, although some people would also use redwood logs if a suitable one washed up onshore. Canoe paddles were carved from hardwoods like Oregon ash, maple, and ocean spray.

Some men were better at the carving arts than others, and usually it was skilled craftsmen who made bows and arrows. Bows for children's practice were made from whale baleen, but most bows were made from vine maple or yew.[23] Craftsmen seasoned the wood, carefully scraped it down, and eventually finished it with a backing of sinew or whale skin. Arrow shafts and points were often carefully made from elderberry or ocean spray.

Basketry was an art practiced mainly by women, although men did work on fish trap baskets. The type of basketry practiced along the Oregon coast was twining—two weavers (sometimes more) woven around a warp. Women began learning as young girls the art of what materials to use, when and where to gather them, and how to season them, process them, and weave baskets with them. Depending on the type of basket, it could be made from sticks of hazel or willow, conifer roots, red cedar bark, maple bark, tule, cattail, or rushes. Decorative overlays and dyes included eelgrass, beargrass, cherry bark, and others. Baskets were used in almost every aspect of life—for packing burdens and for storage, traps, clothing, and mats.

5 Trees

Whenever they lived near the mouth of the river,
in the bay,
they had lots of food.

They had dried salmon,
and likewise (dried) fern roots,
which they ate during the winter.

They ate fern-roots (mostly).
Thus the people did during the winter.

Springbank clover likewise they ate in the winter.
And skunk-cabbage too, was eaten in the winter-time;
also kinnikinnick-berries were eaten.

Such was the food of the people belonging to the past.

—Louisa Smith, Lower Umpqua/Siuslaw[1]

ALDER, RED
Alnus rubra

Hanis: tɬ'wæx
Milluk: tɬ'wæx
Siuslaw: tɬ'waxáim

Plant description: Betulaceae, birch family. Red alder is a deciduous broad-leaved tree that is very common in moist environments along stream banks and floodplains in western Oregon (although in the Willamette Valley the closely related white alder, *Alnus rhombifolia*, is dominant). Mature trees can reach a height of 100 feet; bark is white or light gray; leaves dull green, with double-toothed margins. Flowers are cone-like brown catkins.

Dye: The bark was chewed and spit on basketry materials to create a red dye. It could be used to dye a variety of basketry materials but was used most often on the inner bark of maple (*Acer macrophyllum*) and red cedar (*Thuja plicata*). It could also be used as a red paint on buckskin. Chewing the bark made some people feel ill, and an unidentified plant was chewed as a remedy.[2]

The contemporary method of dyeing is to pound the bark and then mix it with a little water. The dyed material is stirred into the mixture, exposing it to air so that the dye will take.

Games: Alder was sometimes used to make playing sticks for the hand game, known in Hanis as *hǽyæ*. In this game, two teams faced each other. Each team sat behind its own buckskin blanket. There were 100 to 150 playing sticks, which could be made of alder or some other wood like elderberry or ocean spray. These sticks were thin and about 1 foot long. One stick was marked in the middle with paint. One player took the bundle of sticks in his hands. There were also twelve tally sticks, to keep track of points. One player picked up the thin sticks and divided the bundle so that the marked stick was in one of his hands. He then put his hands, holding the sticks, on his thighs while one of his fellow teammates played a hand drum and sang. A player on the other team pointed to the hand where he thought the marked stick was. If the guess was right, the opponent got a turn to hide the sticks. If wrong, the opponent had to pay him one of the twelve tally sticks, which were stuck in the ground alongside the deer-hide blanket. The game ended when one team won all twelve tally sticks. Only men and boys played it, but boys played with a set of fifty sticks.[3]

Among the Upper Coquille, boys would make slides of mud and slide down them on fresh pieces of alder bark with the slick side down.[4]
Other: Alder was one of the nonpitchy woods preferred as firewood in the men's sweat lodge, since it would not produce much smoke.[5] Today, alder is used to cook salmon over a spit, probably reflecting a precontact practice.

ASH
Fraxinus latifolia

Hanis: tłpai
Milluk: tłpa
Siuslaw: čxawəs

Plant description: Oleaceae, ash family. *Fraxinus latifolia* is the only ash tree native to Oregon. They are deciduous and tend to grow in moist areas near stream banks but can also be found in some drier environments. Ash trees can reach a height of approximately 80 feet and a diameter of 3 feet. Their most distinguishing characteristic is the arrangement of their leaves—they are pinnately compound (usually containing five to seven leaflets) and opposite.
Technology: Ash was a prized hardwood widely used to carve canoe paddles.[6] Canoe paddles were about 5 feet long and single bladed, notched at the top or with a small handle tied on with conifer roots or buckskin.[7]

Many Pacific Northwest tribes used Oregon ash wood to make paddles, and sometimes other tools such as digging sticks. Louis Fuller, a Tillamook, thought Oregon ash was superior to any other wood for making canoe paddles, surpassing that of maple.[8]

CASCARA
Rhamnus purshiana

Hanis: wıyípan
Milluk: tum

Plant description: Rhamnaceae, buckthorn family. Cascara is also locally known as chittam. Shrub or small tree, can grow up to 30 feet tall. Leaves deciduous, oblong, with minutely toothed margins and ribbed veins. Bark gray, with white lines and patches.

Dye: Cascara bark was used as a yellow dye for basketry materials (such as the inner bark of bigleaf maple) and as a paint. For paint, the bark (fresh or dry) was pounded and then mixed with a little warm water. Then, using a paint stick, the yellow dye was painted on maple bark headbands or hides.[9]
Medicine: Cascara bark is a powerful laxative. To make a gentler medicine, a decoction of cascara bark and bitter cherry roots was used.[10] Cascara was widely used as a laxative by other Pacific Northwest tribes.[11] Bark extracts have long been used in commercially available laxatives. When I was a child, my family and many of my cousins occasionally peeled and sold cascara bark.
Technology: Cascara was one of the woods carved into shinny clubs. Shinny was a type of indigenous field hockey, a game enjoyed by most tribes in the West. The clubs resembled hockey sticks—straight sticks with a curve at the end.[12] The curve was created by wrapping the end of the stick with marsh grasses and steaming it. The finished stick was oiled with seal or whale oil, to make it glossy and to prevent cracking.[13]

CEDAR, PORT ORFORD — Hanis: láləmɬ

Chamaecyparis lawsoniana

Plant description: Cupressaceae, cypress family. This cedar resembles red cedar but has a white *X* pattern on the bottom of its scalelike leaves, and small round cones. Port Orford cedar, also known as white cedar, grows only along a 200-mile strip of coast from Coos County south to Humboldt County, California.
Technology: Port Orford cedar was sometimes used to make canoes. Its wood is denser than that of red cedar, so Port Orford cedar canoes were somewhat more difficult to haul out of the water. Even so, some carvers preferred working with it.[14]

Unlike other cedars native to the Northwest (Alaska cedar, red cedar, incense cedar), Port Orford cedar has no documented use of its bark or roots in basketry. A Siuslaw man who experimented with it told me that even in young trees, the bark is very thick and not practical to obtain for weaving material. However, the roots of practically any of the region's native conifers can be used in basketry, and it is likely that Port Orford cedar's roots were used in this way to some extent.

In northwest California, the Karuk people used the withes for brooms, and the wood for planks to build sweat lodges, house posts, stools, and headrests.[15]

CEDAR, RED
Thuja plicata

Hanis: tłaháimıł, pgí'ık (roots)
Milluk: tłaháimıł, búwas (roots) (Lower Coquille)
Siuslaw: q'achti

Plant description: Cupressaceae, cypress family. A conifer with scale-like foliage, the underside of which has a white butterfly pattern. Small cones turn upward on the branches. Bark is furrowed and reddish. The tallest trees can reach almost 200 feet. Red cedar prefers moist habitats near the coast but is found as far inland as western Montana.

Fiber: Both the inner bark and roots of red cedar were used in basketry and are still widely used for that purpose today. The inner bark of red cedar was peeled from the tree in spring, when the sap was running. A cut was made through the layers of outer and inner bark with a wedge. Then the bark was peeled in as long a strip as possible going up the tree. The inner bark was separated from the outer bark, and the inner bark was taken home to dry and season. When needed, the bark could be soaked, split into whatever size was required, or pounded to shred it. Cedar bark was used as an overlay to make designs on baskets and was often dyed red with alder bark dye, yellow with cascara bark, or brown with hemlock bark.[16] Three strands of red cedar bark were rolled on the thigh to make rope.[17]

Cedar bark was used to make women's skirts, at least among the Lower Umpqua, Siuslaw, and Alsea (at Coos Bay women preferred to make skirts from the inner bark of maple, as did many people of northwestern California).[18] In 1840 Methodist missionary Gustavus Hines visited a Lower Umpqua village and received some gifts, among them a woman's cedar bark dress, which he described as follows:

> The bark was strung out fine about eighteen inches long, and woven together at one end, so as to admit of being tied around the person, thus constituting a kind of fringe. Two of these fringes made a complete dress; one was fastened around the body above the hips, and hung down to the knees; the other was tied around the neck, and formed a covering for the breast and shoulders; the arms and lower extremities being left perfectly unencumbered. All the women were dressed in this manner.[19]

Roots could be dug at any time of the year. They were processed by heating them up in hot sand and then pulling them through the forks of a fire tong to scrape off the outer bark. Finally, the roots were scraped and split with something sharp, such as a shell. After the roots had been seasoned by drying for several months, they could be used in basketry. Red cedar roots were favored for weaving water storage baskets and cups, as it was thought that they imparted a better flavor to water than other materials. A tightly woven basket of conifer roots can hold water, but to ensure that they were waterproof, these baskets were rubbed on the outside with the roasted roots of an unidentified plant described as having "a tiny potato or wild parsnip like root."[20] Some Rogue River tribes rubbed the insides of baskets with raw camas bulbs to make them more watertight.[21]

Annie Miner Peterson also mentioned that while men were out hunting in the Coast Range, they occasionally brought home roots of another cedar tree that was rare in the Coos Bay region, but was not Port Orford cedar (*Chamaecyparis lawsoniana*). These roots probably belonged to incense cedar (*Calocedrus decurrens*), which can be found in the Coast Range at the eastern edge of Coos County.[22]

Medicine: If a woman did not want to become pregnant ever again, she boiled red cedar boughs, piled them on the ground, and sat on them for as long as they steamed.[23]

Technology: Pounded, shredded cedar bark was one of the materials used for fire kindling.[24] It was also used in a kind of native tinderbox. A coal was wrapped tightly in cedar bark when traveling.[25]

Red cedar is known for its rot-resistant wood and is relatively easy to split and work. Cedar planks were used to build the walls of houses, men's sweat lodges, and storage sheds, and also for roofs on the houses of

Lower Umpqua men in northern-style canoe. Sketch by Captain Albert Lyman, 1850, courtesy of Douglas County Museum.

wealthy people and some sweat lodges (roofs of less elaborate structures tended to be thatched roofs made of a bulrush, *Scirpus microcarpus*). Planks were split from a fallen log or standing tree with the help of elk horn or stone wedges. The planks were about 1 to 2 feet wide, shaped with the help of burning and scraping, and then smoothed by sanding with sharp rocks so that the outer edges were somewhat thinner than the middle, thus helping the planks to overlap in construction.[26]

Planks of fir or cedar were used to make ladders. Holes, flat at the bottom and rounded at the top, were carved into a plank with stone chisels. Short ladders were placed at the entrance of dugout houses to reach the floor. Longer ladders were used to access the rafters of the house, where many foods were stored.[27]

Red cedar was also the favored material for building dugout canoes. Canoe carvers looked for promising fallen logs to build a canoe, but if no suitable logs were found, a tree was felled. This was lengthy and arduous work, done by cutting into a tree all around with wedges and building small fires to burn the base of the tree.[28] A band of wet clay was placed around the tree about 3 feet off of the ground to act as a firebreak.[29]

Alexander W. Chase traveled through the Alsea subagency at Yachats in 1868, working at the time under the Coast and Geodetic Survey. One thing he noted on his trip was the making of canoes:

> These canoes are made by the Indians by selecting a log among the drift timber, so plentiful on all the Oregon beaches; the outside shape is given by the use of the axe, then the interior is carefully hollowed out by using fire and axe. When finished they are from fifteen to twenty feet in length, and the largest will seat from ten to twelve persons. It usually takes an Indian from one to two months' labor, assisted by his wife and family, to make a canoe. Formerly, when they had no tools but stone axes and chisels made of elk horn, it took them sometimes a year to fashion a canoe; but once made they last a lifetime.[30]

Canoe carvers preferred a log at least 3 feet in diameter, if not broader. The log was hollowed out by burning, chipping, and scraping until the general shape was attained. To finish, water and hot rocks were put in the canoe. While the canoe was warm, it was spread wider and crosspieces of a hardwood were inserted to help further shape the canoe and ultimately be the seats in the craft.[31]

Canoes were often finished with red and black paint. Red paint was made by mixing fat with red ochre, black by mixing charcoal with fat. Some canoes were further decorated with inlays of agates or shell on the gunwale, glued in place with pitch.[32]

There were three types of canoes. One was known as the *aludaq*, a word used by Coos, Lower Umpqua, and Siuslaw peoples and borrowed from the Quileute language.[33] These were large, oceangoing canoes with a high prow. In addition to the word *aludaq*, they were also known as *swáhał* in Hanis and Milluk and *síxai* in Siuslaw.[34] The Siuslaw, Lower Umpqua, and Coos Bay people often bought these types of canoes from the Columbia River Chinook or Alsea people, but some were made locally. These canoes were especially good for fishing in the lower bay and offshore in the ocean.

The second type of canoe had rounded ends and was sometimes called a double-ender in English, *máxmax* in Hanis and Milluk, and *łqwá'a* in Siuslaw. This style of canoe was manufactured locally.

Although Chase noted that canoes could hold up to twelve people, some large canoes could hold thirty people. This canoe was believed to be more robust for traveling upriver and navigating among the rocks and rapids.[35]

There was a third type of canoe, a very small one without cross-pieces, known in Hanis as *k'úwuts*, which could hold two people at most. Toy canoes were known by the same name.[36]

Other: There was a grove of red cedars on the south bank of the Coquille River, downstream from the town of Coquille, that the Lower Coquille people named *buwas nıksdída*, meaning "dancing red cedar." The tops of the cedars were bent down and swayed in the wind. Long ago, they were people and had been turned into trees. Their heads were all bent forward when they danced.[37]

Jane Harney, a Hanis woman who lived on the North Fork Siuslaw River after the dissolution of the Alsea subagency in 1876, kept red cedar boughs under beds and transplanted a cedar tree in her yard to keep away fleas.[38]

CHERRY, BITTER
Prunus emarginata

Hanis: dænts
Milluk: dænts

Plant description: Rosaceae, rose family. A tree that can grow up to 50 feet tall. Leaves are deciduous, alternate, 1–3 inches long, with smooth edges. Bark is thin, reddish brown, tends to peel off the trunk in thin horizontal bands.

Fiber: The bark of the bitter cherry was sometimes used as a decorative overlay in basketry. It was also used to make a decorative wrapping on the bottom of large dentalium shells.[39] Dentalium shells were harvested off Vancouver Island and traded throughout North America. In the Pacific Northwest they were used as a currency. It was not unusual for the largest and most valuable shells to be incised with designs.

Medicine: The bark was made into a tea to treat tuberculosis.[40] The roots were boiled with cascara bark to make a laxative.[41]

COTTONWOOD Hanis: bæm
Populus trichocarpa

Plant description: Salicaceae, willow family. Tree, up to 165 feet tall. Deciduous, alternate leaves, broadly triangular, white on the undersides, with smooth or lightly toothed margins. Fruits are round capsules on a string containing cottony seeds.
Other: No cultural uses were noted for this plant. The Hanis name was given by only one informant, Frank Drew. It is similar to the English common names *balm* and *bam*, which are short for balsam, because this tree produces a sweet-smelling resin. It is not clear whether Frank Drew misremembered the Hanis name for this tree, or whether the resemblance is coincidental.[42]

However, this tree had many uses among other Pacific Northwest tribes. Some used the wood to make canoes. Some made ropes from its inner bark. Roots and shoots were utilized for baskets.[43]

CRAB APPLE
Malus fusca

Hanis: sísuxw (green),
mɪč'lǽ'wəs (ripe)
Milluk: sísuxw (green),
mɪč'lǽ'wəs (ripe)
Siuslaw: q'at'i

Plant description: Rosaceae, rose family. Small tree up to 30 feet tall. Leaves ovate to lanceolate, coarsely toothed. Flowers white, occasionally pink. Fruits oval, small, yellowish, occasionally ripening to dark red.
Food: Fully ripened crab apples were simply eaten fresh without any further preparation. Spencer Scott said he ate them as a child, even though they were very sour.[44]

Green crab apples were gathered in late summer after blackberry season. They could be boiled, mashed into a cake, and then dried for winter food stores. They were also mashed with red elderberries with a wooden pestle, and then seal oil and salmon eggs were mixed in. This mixture was eaten by picking up a handful, squeezing it, and licking what came through the fingers. This was done to avoid the seeds of the red elderberries, which contain toxic alkaloids.[45]

Technology: Shinny clubs were made from crab apple wood. The South Slough people favored crab apple shinny clubs.[46] The crook at the end of the stick was shaped by wrapping it with wet grass and steaming it in hot coals.[47]

DOUGLAS-FIR

Pseudotsuga menziesii

Hanis: ságwa (tree), tsgwáhtłas (old-growth tree), mankw (young trees)
Milluk: halq, skwátłəs (bark of old-growth tree)
Siuslaw: łkáihtu (tree), qaułíyu (bark of old-growth tree)

Plant description: Pinaceae, pine family. Conifer with 1-inch-long needles arranged spirally around the branch. Mature trees can grow more than 200 feet tall, with a diameter of 10 feet. Their cones are distinctive, with three-tailed bracts poking out from the scales. Some have described them as resembling small pitchforks.

Fiber: Douglas-fir limbs, called *k'iyas* in Hanis, could be used as the spoke in an openwork pack basket.[48] Occasionally limbs were twined together to make a stiff mat, called *psœ* in Hanis, for drying meat and fish.[49] And Douglas-fir roots were used in basketry as warp or weft, anywhere spruce roots were used.[50]

These *psœ* were also used to create a kind of caravan for fish camps. In autumn, people from coastal villages moved to upriver camps to catch migrating salmon and lamprey. At the end of the season, three canoes were connected with the *psœ*, with the outer canoes guided by three paddlers, the middle canoe by only two. On returning to the downriver villages, the *psœ* was stored in the rafters of the houses.[51]

Douglas-fir limbs and roots were important in the manufacture of large fish trap baskets, known as *hak* in Hanis, *kəmatłáts* or *tsú'ʊn* in Siuslaw. These were large conical baskets up to 14 feet long, woven around vine maple hoops.[52] The sticks were twined together with conifer roots, usually of Douglas-fir or spruce. Cedar roots were not usually used on fish trap baskets, as it was believed they had too strong an odor, unlike Douglas-fir or spruce roots.[53] These baskets were placed at junctions in fish weirs, which were fences built in the estuaries to trap fish. The baskets had trap doors on top so fishermen could remove fish without removing the whole trap basket.

The roots were used as warp and weft in other kinds of baskets, and for making handles.[54]

Food: Small pieces of Douglas-fir boughs were steeped in warm water to make a tea. The water was not allowed to boil, and the tea was drunk lukewarm.[55]

Medicine: The tea made of steeped Douglas-fir boughs was also used as a healing wash. Indian doctors used it to wash people's faces after they attended a funeral.[56]

The house of a deceased person was also cleaned with Douglas-fir limbs. First, the limbs were burned throughout the length of the house to thoroughly fumigate it. Then an Indian doctor washed the walls and ceiling with Douglas-fir limbs dipped in water.[57]

Other: Douglas-fir limbs were laid into graves, below and along the sides of coffins. Often the top board of a coffin was made of Douglas-fir wood. It was believed that the presence of Douglas-fir would slow the process of decay.[58]

A fence of Douglas-fir, with grass stuffed between the closely set poles, was used for privacy around swimming holes where women went to bathe.[59]

Women gathered the thick bark of old-growth Douglas-fir, with the help of elk horn wedges, for use as firewood.[60] The bark was said to burn slowly and give more heat, so it was the preferred fuel for hearth fires in the plank houses.[61]

Sometimes people fished at night for sturgeon on Coos Bay. Platforms were put on a canoe and covered with sand, and a fire of Douglas-fir or pitchy woods was lit on top of that to give a good bright light.[62]

There was a local variation of the game of tossles (sometimes called double ball), known in Hanis has *naldáał.* Among the Coos, Lower Umpqua, and Siuslaw peoples, it was played only by women. The tossles were two short pieces of wood tied together with leather or perhaps cordage. The throwing stick was 3 to 5 feet long and often carved from

Douglas-fir wood. Women would pair up against each other and throw the tossles beyond the goal, set about 100 feet away from the players. Five or six goals made a game, and the players made small wagers on the outcome.[63]

Technology: Douglas-fir poles, and sometimes large roots, were placed in the mudflats to form fish weirs (called *tłəm* in Hanis and Milluk, *mati* in Siuslaw).[64] The poles were lashed together with cross posts and conifer root ties.[65] People also used trap baskets, nets, and fishing spears in conjunction with weirs to catch numerous species of fish—herring, flounder, suckers, salmon, lamprey (colloquially known as eels), and many other species. Smaller weirs and trap baskets constructed upriver caught migrating salmon and lamprey.[66]

The shafts of fish spears were made from young Douglas-fir trees. They were whittled and then smoothed with horsetail, which was full of silica and was often used for polishing. The shafts were usually about 10 to 14 feet long, with spear points of bone or hardwood attached to them.[67] The shaft of a fishing spear was often made with one side flattened, so the fisherman could tell which way the hook was turned. Fishing spears were often used for fishing from a canoe at night or from brush fishing shelters built along the riverbank.[68] The spear was held vertically with the points resting against the bottom of the bay or river. When the fisherman felt a salmon touch it, he jerked the pole up.[69] Douglas-fir poles were also used to make the shaft for dip nets.[70]

Douglas-fir was occasionally used to make canoe paddles, although hardwoods such as maple and ash were generally preferred.[71]

Douglas-fir was often used as support posts for houses and storage sheds.[72]

For building meat-drying racks, whittled old-growth Douglas-fir was preferred because it split well and had no pitch. Wood from young trees could be pitchy and might impart off flavors to fish and meat on the drying rack. For the same reason, cedar was never used to build drying racks.[73]

Douglas-fir skewers and split sticks were used to roast salmon or slabs of meat.[74] A stick could also be run through the gills of smelt or through mussels and hung over a smoky fire for curing.[75]

The least valuable arrows were whittled from Douglas-fir sticks. In Hanis this type of arrow was called *tłə́bəč'*. The pointed end was made to be slightly heavier, with a small notch in the other end. These arrows were used by adults for practice shooting and by boys for hunting ducks.[76]

FIR, GRAND Hanis: mankw

Abies grandis

Plant description: Pinaceae, pine family. The grand fir is the only true fir (conifers in the genus *Abies*) that grows near the Oregon coast. The tops of its needles are bright green and the bottoms have a whitish bloom. Needles tend to grow in two flat rows along the branches. Cones grow upright from the branches.

Other: No cultural use for grand fir was given. However, its roots were probably used occasionally as basketry or ties, as were those of Douglas-fir and other conifers, and the wood may have been used in fish weirs and meat-drying racks.

According to Frank Drew, second-growth Douglas-fir was called *mankw* too, so perhaps both young Douglas-firs and grand firs were known by the same name in Hanis.[77]

HEMLOCK, WESTERN Hanis: č'ǽmæł, qbahq

Tsuga heterophylla Milluk: č'ǽmæł

Plant description: Pinaceae, pine family. Western hemlock has the shortest needles of any of the conifers native to western Oregon, less than 1 inch long, and each twig is full of needles of variable lengths (hence the species epithet *heterophylla,* meaning "variable leaf"). There are two white lines on the underside of each needle. Cones are quite small, no more than 1 inch long. These trees are distinctive at a distance because their branch tips and treetops are droopy.

Dye: Hemlock bark was used to make a brown dye. The bark was boiled and conifer roots soaked in the water overnight. Fishnets were sometimes dyed brown to make them harder for fish to see.[78]

Other: Jim Buchanan recalled a myth of a hemlock who was a person. There was a man wind who was powerful and broke trees, and a woman wind who uprooted trees with her digging stick.[79]
Technology: Hemlock was one of the woods used to build fish weirs. It was dense and would not float away.[80]

If a young hemlock tree had a crook in it, it was good material for making a shinny stick.[81]

MAPLE, BIGLEAF
Acer macrophyllum

Hanis: húlık'
Milluk: húlık'
Siuslaw: sna

Plant description: Aceraceae, maple family. The bigleaf maple is aptly named—its five-lobed leaves can grow to be 1 foot across. A mature tree can grow up to 160 feet tall, although heights of 50 to 70 feet are more typical. The grayish bark of older trees can be completely hidden under a thick cover of mosses and ferns.
Fiber: The inner bark of maple was used occasionally in basketry. It was gathered in spring. Its primary use as a fiber at Coos Bay was to make clothing. The bark was pounded with a wooden beater over a log and then tied from a buckskin belt or twined at the top to make a skirt, in the same style as the cedar bark skirts made by the Lower Umpqua, Siuslaw, and Alsea. To make the skirts more decorative, some bark strips were dyed red with alder bark, and some were buried in a black mud to turn them black; these dyed strips were then alternated with plain bark to make a striped skirt. Each colored stripe was made of about ten or so strips of bark.[82]

Strips of finely pounded maple bark were braided into headbands, worn by children and women (men tended to wear buckskin headbands more than those of bark). Women and girls further decorated their headbands by using finely pounded bark to make "pom-poms," attached to the sides of the head.[83]

Women made capes and skirts of maple leaves or sword ferns for the annual solstice dance, known as the time when "the sun goes back."[84]

A large piece of maple bark was folded into a trough for cooking food.[85]

Technology: Maple was one of the local hardwoods carved into canoe paddles.[86] The wood was also carved into pestles for pounding bark or foods such as berries.[87]

MYRTLEWOOD
Umbellularia californica

Hanis: šíčıls (nuts), wægænł (tree)
Milluk: šíčıls (nuts)

Plant description: Lauraceae, laurel family. In Oregon, this tree is usually called myrtlewood, but in California it is more commonly known as the California laurel or bay laurel. Myrtlewood grows in the southwestern corner of Oregon, as far north as the Umpqua divide, and throughout much of California. Mature trees can reach heights of 80 feet or more and often have multiple trunks. Leaves are dark green, up to 6 inches long, and arranged alternately. Flowers are pale yellow and small. In autumn the flowers develop into yellow-green fruits about the size of a large olive. When ripe they turn dark purple.
Food: The nuts were gathered in the fall, hulled, and dried. They could be stored in baskets for winter fare and were prepared by roasting them over hot ashes; they were then eaten whole, often with salmon eggs.

Myrtlewood, *Umbellularia californica.*

Raw nuts were bitter.[88] However, when properly roasted they had the interesting flavors of coffee, bitter chocolate, and burned popcorn.

Myrtlewood nuts were an important food for many native peoples in southwestern Oregon and California. David Douglas, the famed Scottish botanist who made a foray into the Umpqua Valley in 1827, noted that fur trappers and hunters made a tea from the bark of this tree.[89] The Upper Umpqua Indians ate the roasted nuts of this tree and made a tea from leaves and young shoots that in his opinion was "by no means an unpalatable beverage."[90]

Medicine: There is a Coquille tradition that myrtlewood leaf tea could be used to draw boils.[91]

OAK

Quercus garryana, Q. kelloggii, Q. chrysolepis, Notholithocarpus densiflorus

Hanis: šɪšda (tree), álam (acorn)
Milluk: álám (acorn)
Siuslaw: múxwa (tree), qwna'ax (acorn)

Plant description: Fagaceae, oak family. California black oak (*Quercus kelloggii*) grows in the Umpqua and Rogue Valleys and in California. These deciduous trees have pinnately lobed leaves, with each of the seven lobes pointed on the end. Acorns are over 1 inch long and have deep caps, and they are more oval and pointed on the tips than those of white oak.

Oregon white oak (*Q. garryana*) has leaves that resemble those of black oak but with rounded tips. White oak acorns are less than 1 inch long and somewhat round, with shallow caps. White oak grows in the Willamette Valley as well as in the Umpqua and Rogue Valleys.

Canyon live oak (*Q. chrysolepis*) grows in the Umpqua and Rogue Valleys and can be found near the coast in southern Curry County. In dry country it grows as a bush or small tree, but in wetter habitats it can grow up to 60 feet tall. Unlike white oak and black oak, this tree is evergreen. The leaves are dark and shiny, and although some are smooth edged, most have sharp spines along the margins and resemble holly leaves. The oval acorns are 2 inches long and have shallow caps.

Tanoak (*Notholithocarpus densiflorus*) grows in Curry County south to California's Bay Area, with a few scattered populations in the Sierras.

Acorns are oval, with spike-covered caps. Leaves are thick and leathery, dark green, up to 5 inches long, and often have hairs on the underside. In the mountains tanoak grows in a shrub-like form but in the coastal fog zone it can grow as tall as 100 feet.
Dye: Oak bark, like hemlock bark, makes a brown dye.[92]
Food: Oak trees grew in the inland valleys and in Curry County, outside the territories of the Coos, Lower Umpqua, and Siuslaw, who traded with their inland and southern neighbors for acorns. Acorns contain bitter tannins, which could be leached by burying them in the ground, or by pounding them and rinsing the meal with water. Once the acorn meal was ready, it was boiled in cooking baskets and eaten.[93]

Coquelle Thompson said the Upper Coquille people burned oak prairies every fifth year in late summer.[94]
Other: The Hanis term *šíšda* seems to be borrowed from Upper Coquille, an Athabaskan language. The Upper Coquille word for oak is *šášda'*, and the Coos Bay people traded for many of their acorns with them.[95] Coos people also called oaks "acorn's tree"; in Hanis *ha'álamu nɪk'ɪn*, and in Milluk *álam dɪ nɪk'ɪn*.[96]

There is some confusion over the term *qwna'ax*—in one source this word is listed as the Siuslaw word for acorn, but elsewhere it is listed as the word for hazel nut.[97]

PINE, GRAY Hanis: mái'nus (nut)
Pinus sabiniana Milluk: máinus (nut)

Plant description: Pinaceae, pine family. This species grows primarily in the dry hills ringing the Central Valley of California, although a couple of small populations have been documented in southern Oregon's Jackson County. Today few gray pines are taller than 75 feet, although two hundred years ago the tallest were reported to be 100 feet tall. Needles are 1 foot long, arranged in bundles of three, and grayish in color. The cones are large (up to 15 inches long) and have spurs on the end of the scales.
Other: Indians in California ate the seeds and made beads from the nuts of this pine, and through trade these were acquired by the Coos, Lower Umpqua, and Siuslaw peoples. The pine nut beads were strung

Indian Kate, circa 1900, wearing pine nut apron. Photo courtesy of Coos Historical and Maritime Museum, 995-D184.

on necklaces or on a belt to create a decorative apron for women to put over their dresses at dances.[98]

Pine nut beads in their natural state are brown. Cooking them in oil turns them a shiny black.[99]

PINE, SHORE
Pinus contorta var. *contorta*

Hanis: cɪpk
Siuslaw: yákwɪm, mənkw

Plant description: Pinaceae, pine family. Shore pine is easy to identify because it is the only indigenous true pine on the Oregon coast. Its leaves are in bundles of two, and it grows within a narrow band along the coastline. The trees tend to be small and shrub-like. This

same species (although a different variety, *P. contorta* var. *latifolia*) is the lodgepole pine found in the Cascade Range and elsewhere.
Fiber: The roots were used as strong ties for bundles of dried salmon. Sometimes the roots were incorporated into baskets, along with other conifer roots.[100]
Other: Shore pine, along with spruce and Douglas-fir, was one of the main sources of pitch, *c'ә́łan* in Siuslaw, *satł'* in Hanis and Milluk. A little pitch was used to secure a quill to arrow shafts, and sinew to fishing poles. Pitch was used to patch leaks in canoes by heating it, mixing it with seal oil, and rubbing it into the leak with a hot rock.[101]

REDWOOD Hanis: mı́kit
Sequoia sempervirens

Plant description: Redwoods are the tallest trees of the Pacific coast; the tallest reach over 300 feet. Their natural range is western California from Santa Cruz north to the Chetco River.
Technology: Although the northern edge of the redwood forests is about 100 miles south of Coos Bay, redwood logs washed up all along the coast. Some tribes, like the Tillamook of Oregon's north coast, utilized these logs to make canoes.[102]

However, the Milluk people of South Slough considered some large washed-up logs off limits. Nellie Wasson Freeman (Milluk/Upper Coquille) told the story of a Chetco man, Silas Tichenor, living at Coos Bay, who violated this taboo:

> A big redwood log was washed ashore on Merchant's Beach. The [South Slough] Indians would never touch it. It must have been brought by the big tidal wave they talk about. They thought it had been sent by some great being for a purpose and it lay there for centuries never rotting. The Indians reverend [*sic*] it. In later years, in our days a Chetkoe [*sic*] Indian named Silas came here and he made a canoe from the log. The Indians were shocked. Mama said "Something is going to happen to Silas. He had no business to disturb the trees put there by greater hands." He made the canoe and worse still he brought it to South Slough over

the bar. The [South Slough] Indians wouldn't touch it. . . . After Silas left the Indians took the canoe and put it way up a creek. We called it Canoe Cove. After Silas died I heard that Ione [Silas's sister] had it taken over the bar and sunk. Silas died in 1887.[103]

SPRUCE, SITKA
Picea sitchensis

Hanis: č'ɪšíməɬ, č'ɪšímɬ
Milluk: č'ɪšímɬ
Siuslaw: c'así

Plant description: Pinaceae, pine family. Sikta spruce is found only within the coastal fog belt. Mature trees can be 200 feet tall. The needles are about 1 inch long, with sharp tips. Bark is broken up into small scales. Pollen cones are red; seed cones are 2 inches long, with very thin scales.
Fiber: Switches of spruce made a sturdy warp for pack baskets.[104] Limbs could also be woven into wooden dippers.[105]

Spruce roots were utilized as warp and weft material in a variety of baskets: berry baskets, fish traps, grinding baskets, pack baskets, plates, or any basket that needed strength and durability. Spruce roots also made good ties for dip nets and bundles of dried meat, and they were even used to make earrings and nose pendants.[106] The roots were heated to remove the bark, and the root could then be split down to whatever size was required.[107]
Other: A child's top, called *tɬáɬti* in Hanis, was made from a disk of spruce bark with a small hole in the center for a fir stick. A child would spin the stick between the hands, making the spruce disk spin.[108]

Young spruce trees were sometimes used, along with vine maple, to make the hoops in the hoop-and-pole game *tɬaxaúk'wanawas*. For a description of the game, see the entry under red elderberry.[109]
Technology: Spruce was one of several woods that were carved and polished into bowls and cups. A bulge was removed from the tree with chisels. Then hot coals were placed in the center, and the burned wood scraped away. The carver knocked off charred wood from the sides and bottom, making sure that the bowl was not burned too deeply in any one place. Then, the bowl or cup was polished with a rough stone, inside and out, with perhaps a little final polishing with horsetails.[110]

Occasionally, small shovel-nosed canoes were carved from spruce.[111]

YEW Hanis: kásai
Taxus brevifolia

Plant description: Taxaceae, yew family. This is a slow-growing, small tree (rarely taller than 45 feet) in the forest understory. Bark is gray on the outside and purplish on the inside, and often falls off the tree in thin sheets. Needles are two ranked and flat. These trees do not produce cones like other conifers. Female trees produce seeds in an "aril" that resembles a red huckleberry but is poisonous to humans.

Technology: Yew wood was used to make bows. The wood was seasoned, carefully carved, and backed with sinew or whale skin.[112]

Hoxie Simmons, a Rogue River man, said the wood for a bow should come from the side of the tree without knots, usually the east- or south-facing side. The north-facing part of the tree was more likely to have knots.[113]

Yew or ocean spray was also used to make the detachable end of harpoons.[114]

6 Shrubs

BLACKBERRY
Rubus ursinus

Hanis: wixaini, wíxáinı
Milluk: dzudzua
Siuslaw: c'xát'aat'

Plant description: Rosaceae, rose family. Trailing, prickly vines, up to 25 feet long, typically found in burned or logged areas. Vines are dioecious (either male or female); only female vines produce berries. Leaves are alternate and slightly hairy; flowers are small and white; berries ripen in July.

Blackberry, *Rubus ursinus*.

Food: This trailing blackberry was, and still is, a much-loved food. The berries ripen in July and the native people traveled far up into the hills to pick them. Fresh berries were a treat, but many were dried for winter food stores. Mats (made of tule, fir sticks, or hides) were stretched out on the ground, and the berries were spread on them to dry in the sun. The dried berries were stored in baskets in the house for winter food. Some berries were pressed into cakes, sometimes also mixed with other berries and fruits such as crab apples, and then dried.[1]

One Lower Umpqua elder said that sometimes his grandmother crushed blackberries and put the juice on salmon before smoking the fish.[2]

If old vines were no longer producing berries, fern fronds were broken off and a thick layer of them placed under the barren vines. They were then supposed to bear fruit again.[3]

Dried blackberries were enjoyed by soaking them in water before eating. Sometimes a beverage was made by allowing the juice to ferment. It was called *lə'əl* in Hanis and only men drank it.[4]

The vine and leaves of the plant were also gathered to make tea, called *yaxdanáʼał* in Hanis (which in one source was also glossed as the word for blackberry vines). The leaves were steeped in warm water in a basket near the hearth fire.[5]

Two other species of blackberries were introduced in the late nineteenth century to North America from Eurasia, the Himalayan blackberry (*Rubus bifrons, R. discolor*) and the evergreen blackberry (*R. laciniatus*). They have both become common in the wild throughout many parts of the Pacific Northwest, to the point of being invasive. The native people recognized them as a recent introduction and named them in Chinook jargon the *bastan úlali*, literally "the white man's berry."[6]

BLACK CAPS
Rubus leucodermis

Hanis: dlæpsǽn
Milluk: dlæpsǽn

Plant description: Rosaceae, rose family. Shrub up to 6 feet tall, prickly stems covered with whitish bloom; leaves alternate, deciduous, usually with three sharp-toothed leaflets with white undersides. Flowers small and white. Fruit initially red, ripening to purple or black.

Food: Black caps are a species of raspberry. They were ripe at about the same time as blackberries. They were enjoyed fresh or dried for winter storage.[7]

BLUEBERRY
Vaccinium uliginosum

Hanis: q'áni
Milluk: q'áni

Plant description: Ericaceae, heath family. Low-growing shrub, rarely more than 2 feet tall, in swamps and wet meadows from the coast to the west slopes of the Cascades. Leaves deciduous, alternate, and oval. Flowers small, pinkish, and urn shaped.
Food: This species of blueberry (often called bog blueberry, because like many other members of the heath family it grows in boggy habitats) ripens in September, about the same time as the black huckleberry (*V. ovatum*). The native people picked bog blueberries and ate them fresh, but they did not eat nearly as many of them as they did huckleberries, as they were not regarded as tasty.[8]

When glass beads were introduced, the Coos Bay Indians nicknamed blue beads *q'ani*, blueberry.[9]

CURRANT, RED FLOWERING
Ribes sanguineum

Hanis: tɪptíplɪ
Milluk: tɪptíplɪ

Plant description: Grossulariaceae, gooseberry family. Branching shrub, 3 to 6 feet tall. Leaves with three to five lobes, somewhat serrate. Blooms from late March to May, flowers in a raceme of pink-red blossoms. Fruit black, flavor often described as insipid.
Other: The berries of red-flowering currant are edible but not particularly flavorful, and they were generally not eaten. In a myth, the Trickster used the beautiful red blossoms of the red-flowering currant as imitation red-headed woodpecker scalps and masqueraded as a wealthy man.[10]

This term was also used for other closely related species of currants, with white flowers. None of these species were generally eaten either. Lottie Evanoff observed that even bears would not eat these berries.[11]

ELDERBERRY, BLUE
Sambucus nigra ssp. *caerulea*

Hanis: líšwat

Plant description: Also identified as *Sambucus caerulea*, part of the family Adoxaceae, formerly part of Caprifoliaceae, the honeysuckle family. Blue elderberries are shrubs, sometimes more than 30 feet tall, with pinnately compound leaves consisting of five to seven leaflets with toothed margins. Flowers are creamy white, blossoming in June and July in a flat-topped inflorescence.
Food: Blue elderberries begin ripening in August, although they keep ripening through October. They rarely grow near the coast and so were not gathered as frequently as red elderberries. Blue elderberries were often eaten by mixing them with crab apples.[12]

ELDERBERRY, RED
Sambucus racemosa var. *arborescens*

Hanis: mahá'wai
Milluk: txai

Plant description: Family Adoxaceae, formerly classified as part of Caprifoliaceae, honeysuckle family. Red elderberries are shrubs, sometimes more than 30 feet tall, with pinnately compound leaves consisting of five to seven leaflets, with toothed margins. They are distinguished from blue elderberries by pyramid-shaped flower clusters (rather than flat-topped), and their ripe berries are red. Blue and red elderberries can be found growing in the same region somewhat inland from the coast, but in the coastal zone only red elderberry is found.
Food: Red elderberries ripen in early summer. The berries were mashed in a wooden bowl, and the pulp mixed with salmon eggs. Later in the season, crab apple and seal oil were added to the dish as well. These mixtures were squeezed and licked from the fingers. The seeds were always carefully discarded, as they contain toxic alkaloids.[13]

One Lower Umpqua man commented that red elderberries were more bitter than blue ones.[14]
Other: Long sticks of elderberries, about 7 to 8 feet in length, were cut to make throwing poles in a hoop-and-pole game, known in Hanis as

Red elderberry, *Sambucus racemosa.*

tlaxaúk'wanawas, which men played whenever the weather was good. The elderberry sticks were put up in the house rafters to dry and season. Then the poles were tapered at one end, smoothed and straightened, oiled, and wrapped with sinew and pitch near both ends to prevent splitting. Long poles of hazel were bent so each end was set in the ground, perhaps 40 to 50 feet apart from each other. About 30 feet away from the hazel hoops, goalposts were placed at either end. A player would run to the first goalpost and throw the pole, trying to send it over the two hazel poles and past the other goalpost. For every successful throw, a player earned a point. Points were kept track of with tally sticks made from ocean spray (*Holodiscus discolor*).[15]

Children also played their own game of toughness called the burning game; *xábəp'áa'nɪ'was* in Milluk, *xabə́p'a'náwas* in Hanis. Small elderberry sticks were gathered in autumn and dried by the fire. The sticks were cut into pieces 1 to 2 inches in length. Children would bet small wagers on who could keep burning sticks on the back of their hand for the longest time. Some played by burning two or three punks at once on the back of the hand.[16]

Technology: Rattles, about 1.5 feet long, were made from elderberry branches that were whittled and split so that one side was thinner and more limber, and the other stiffer, so the thinner side would smack against it. Sinew was wrapped around them to keep them from splitting too far. The rattles were part of the ghost dance, introduced to western Oregon in the 1870s, although these rattles may have been used along the Oregon coast before that.[17]

One type of arrow shaft was made from red elderberry wood. In Hanis, the plant as a whole was called *mahá́wai*, but the wood in the main trunk was called *púkwœ*. Most arrow shafts were made and sold by skilled craftsmen. Numerous branches were cut from the shrub, about the thickness of a finger. The bark was scraped off, and the sticks bundled and hung up from the rafters of a house for a few days to dry. Then the bundle was taken down and each stick was smoothed and straightened by working it through a groove in a flat rock, already warmed next to the fire.[18]

Projectile points made of *Holodiscus discolor* were often used with elderberry shafts. The heartwood of elderberry is easily hollowed out, and an ocean spray point could be inserted into the hole and secured with a wrapping of eagle feather quills and salmon skin glue or pitch.[19]

The arrows were fletched with eagle feathers and wrapped with sinew and salmon skin glue. Finally, each arrow was decorated with paint, usually red paint that was derived from red ochre mixed with elk marrow. Rarely, blue paint was used, but there were few sources of it. One place the blue clay paint was gathered was at Cook's Chasm near Cape Perpetua.[20]

Some Oregon coastal peoples made stems for smoking pipes from red elderberry.[21]

GOOSEBERRY

Ribes divaricatum, R. menziesii, R. bracteosum, R. laxiflorum — Hanis: tax'wái, dılxáxa

---

Plant description: Grossulariaceae, gooseberry family. For berry pickers, the basic difference between gooseberries and currants is that gooseberries are usually prickly, and currants usually not. The flowers tend to be different also—gooseberry flowers tend to point down to the ground and the stamens project noticeably past the petals, which is usually not the case for currants.

All are deciduous shrubs, with leaves that vaguely resemble those of maples, and most can be found in moist habitats in full sun or partial shade. All species of *Ribes* found in the Coos, Lower Umpqua, and Siuslaw region have blue or black berries.

Coast black gooseberry (*R. divaricatum*) ripens the earliest of any of the local *Ribes*, in July and August. The other varieties ripen a bit later.

Food: There are several species of currants and gooseberries in the genus *Ribes* in western Oregon. Annie Miner Peterson mentioned two species of gooseberry that were eaten fresh in season. *Tax'wái* was a gooseberry that was ripe in fall, and *dılxáxa* was a current or gooseberry with black fruit that was eaten in summer. Given that coast black gooseberry ripens earlier than the other varieties, it may be *dılxáxa.*[22]

Lottie Evanoff said there was a type of climbing currant with tiny white flowers, and its berries were so awful that not even bears would eat them. This was probably the trailing currant, *R. laxiflorum.*[23]

HAZEL
Corylus cornuta var. *californica*

Hanis: wíyæ (bush), tæ'læ'məs (nut)
Milluk: tsʊst (bush), tæ'læ'məs (nut)
Siuslaw: čistx

Plant description: Betulaceae, birch family. Deciduous shrub 3 to 12 feet tall, leaves with doubly saw-toothed margins and a slightly heart-shaped base. In spring hazels flower before growing new leaves. Male flowers are in catkins, and female flowers are in very small catkins. The nuts are round, with a hairy covering and a long "beak" that protrudes past the nuts. The nuts appear in clusters of two or three at the ends of the branches.

Fiber: Hazel switches were used as a warp in many baskets. Switches were gathered in spring, the bark peeled, and the sticks dried and stored in a shed to be ready for use. Twined with spruce roots, they were made into large openwork pack baskets.[24] They were also gathered when out elk hunting and woven with willow bark to make a basket to pack elk meat back home.[25] Fine glossy sticks were woven into a small ark-shaped basket, called *bú'us* in Hanis, which was used to store valuables like dentalium shells.[26]

Conical fish trap baskets (*glúbət* in Hanis, *tsu'wun* in Siuslaw) were made with hazel sticks and conifer roots for use in upriver lamprey fishing sites. These baskets were about half the size of the large trap baskets in the estuaries.[27]

A small-handled dipper (*díbu* in Hanis and Milluk) made of wood, hazel switches, or spruce limbs was used to dip small fish into a basket. It was always kept in a canoe. Dippers carved from solid wood were also useful for bailing water from a canoe.[28]

In northwestern California and southwestern Oregon, people made basket baby cradles from hazel sticks. The Coos, Lower Umpqua, and Siuslaw made some cradleboards in this style, but most cradles were made with a flat fir or cedar board for a back, and buckskin wrapping and vine maple hoops.[29]

Food: Hazel nuts were harvested in autumn and were eaten either raw or roasted. Hazel nuts were stored in the house in bags or baskets for winter.[30]

Other: Long hazel sticks were used to make poles for the hoop-and-pole game. For a description of the game, see the entry under red elderberry.[31]

Groves of hazel were maintained by burning patches approximately every five years. Burning made the trees grow back with long, straight shoots for basketry and a reinvigorated crop of nuts.[32]

Technology: Hazel poles, at least 12 feet long, were cut and carefully seasoned by heating over a fire. These strong but flexible poles (*háutłau* in Hanis, *híq'ai* in Siuslaw, *yúq'wi* in the Lower Umpqua dialect) were used for poling canoes in rapids in the shallows upriver.[33]

HUCKLEBERRY, BLACK
Vaccinium ovatum

Hanis: q'áxas (black), pasásıya'wa (blue)
Milluk: q'as
Siuslaw: táxxai, č'eixan

Plant description: Ericaceae, heath family. Bushy shrubs than can grow up to 12 feet tall. Leaves 1 to 2 inches long, leathery and dark green on

Black huckleberry, *Vaccinium ovatum*.

top (sometimes with some dark red along the edges), with saw-toothed edges. Flowers pinkish, resemble small urns; berries shiny black or powdery blue, ripen in late summer and early autumn, but can sometimes still be picked until early November. Black huckleberries grow in the understory of coastal forests and on sand dunes.

Food: The berries ripen in September. Many were eaten fresh, and many more were dried for winter food stores.

People observed that some bushes had shiny black berries, and other bushes produced berries that looked powdery blue. It was thought that the bluish berries were slightly larger and sweeter than the black ones, but only the Hanis people gave different names to the two types of black huckleberry.[34]

Other: Shinny balls were carved from the roots of black huckleberries (as well as the roots of two other members of the heath family, rhododendron and kinnikinnick). These small balls, about 2 inches in diameter, were whittled, smoothed, and oiled with seal or whale oil.[35]

HUCKLEBERRY, RED
Vaccinium parvifolium

Hanis: tłæxtłǽuxæs
Milluk: tłæxtłǽuxæs
Siuslaw: ya'úwɪ, ya'úha

Plant description: Ericaceae, heath family. Grows as a shrub up to 12 feet tall in partial shade or full sun. The twigs are bright green, and the leaves are light green (especially compared to other members of the genus *Vaccinium*) and small (about 1 inch), with smooth edges. Flowers resemble those of black huckleberries; berries are red and ripen in midsummer.

Food: The Hanis and Milluk name for red huckleberry means "whip down," describing the common method of gathering these berries—by spreading blankets below the bushes and beating the berries from them.

A word of caution came with gathering these berries. It was said that if a person picked the berries until evening, she would see more and more bushes and feel compelled to keep picking, going farther and farther into the woods. Then she would get lost and become an *œšən*, wild being of the woods. So berry pickers were urged to stop picking red huckleberries in the early afternoon and head home.[36]

Other Oregon coast peoples also associated red huckleberries with dangerous beings. Among the Tillamook, red huckleberries belonged to Wild Woman. People were not supposed to eat these berries out in the woods, or Wild Woman might look at them and make them insane. Also, red huckleberries were one of the foods that a young woman was forbidden from eating at all in the year after her first menses.[37] Among the Alsea, these berries were associated with the blood of *Asɪn*, the wild woman, and so there were taboos about who could eat the berries, and when. Among the Alsea, young women during their first menstruation ate only dried foods, and the only berries they could eat were red huckleberries.[38]

Other: Shinny balls could be carved from the roots of the red huckleberry plant, but they were more commonly made from black huckleberry roots.[39]

KINNIKINNICK
Arctostaphylos uva-ursi

Hanis: báhwiya, báxwya
Milluk: báhwya
Siuslaw: lállap (plant), p'íyʊxun (berries)

Plant description: Ericaceae, heath family. Ground-hugging shrub that often forms mats. Leaves are oval, about 1 inch long, alternate. Flowers are pinkish and urn shaped. Berries are red, ripen in late summer, and remain on the shrub through the winter.

Food: Kinnikinnick berries were eaten, although they were not as favored as other berries such as blackberries and huckleberries. They have a mealy texture and not much flavor. The berries were prepared in a couple of different ways. One was to pound them fine and eat them raw. Or they were eaten whole and mixed with salmon eggs. Sometimes the whole berries were heated with salmon eggs in a flat pan made of spruce roots that was jiggled high enough over a fire to keep the pan from burning through.[40]

When cranberries were introduced, they were called by the same names as kinnikinnick.[41]

Other: Leaves were toasted over a fire and mixed with tobacco (*Nicotiana quadrivalvis*) for smoking. The leaves had a pleasant smell and improved the taste of the tobacco. The best leaves for smoking were those of plants that grew in places protected from the wind. Lottie Evanoff remembered that when her family traveled, her father, Doloos Jackson, especially liked the kinnikinnick growing in the dunes near Clear Lake, just south of the Umpqua River, for smoking.[42]

Beverly Ward recalled helping her husband's Lower Coquille grandmother, Susan Ned, gather kinnikinnick leaves: "Sometimes we helped Grandmother gather kinnikinnick on the sandhills and she supervised the operations and every leaf had to be just so. She said some Indians smoked plantain and broad-leaf dock, but she liked kinnikinnick the best. She mixed a little tobacco with the dried leaves to smoke in her pipe."[43]

In an Alsea legend that was probably known to the Siuslaw and Lower Umpqua, the world transfomer *S'úku* began the custom of mixing kinnikinnick and tobacco. As he stood by the Umpqua River, preparing to catch salmon, he called up a pipe and tobacco. He dropped some tobacco into the kinnikinnick growing there. For this reason, the people who were yet to come would always smoke kinnikinnick and tobacco together.[44]

Children took the berries and strung them as make-believe money beads, along with stems of dried clover and wild lettuce.[45]

LABRADOR TEA

Rhododendron columbianum

Plant description: Labrador tea (formerly *Ledum glandulosum*) is in the heath family, Ericaceae. Small shrub, usually no more than 3 feet tall, grows in boggy habitats. Leaves elliptical, with rounded tips, droop slightly downward from the stem. Blooms in early spring, with white flowers.

Food: The leaves of this plant were gathered to make tea, and it is still a popular wild tea today. Historically, the tea was also popular throughout the Pacific Northwest, with pioneers and Native Americans alike.[46]

Labrador tea, *Rhododendron columbianum*.

One elder recalled that they picked Labrador tea in October and let it dry in bags. The dried leaves were soaked in hot water to make tea, which was supposed to be good for the kidneys.[47]

There is a Coquille tradition that the tea is good for menstrual cramps.[48]

MANZANITA, HAIRY	Hanis: bi
Arctostaphylos columbiana	Milluk: bi

Plant description: Ericaceae, heath family. Mature shrubs 3 to 10 feet tall, leaves small (1 to 2 inches long), covered in fine hairs and arranged alternately. The urn-shaped flowers are white to pinkish. Berries when ripe are small and brownish red.

Food: The berries were picked in the fall, about September. They were pounded into a fine flour in a *mádan*, a grinding basket. Then this manzanita flour was mixed with dried or fresh salmon eggs into a kind of mush or cake. Salmon egg–manzanita cakes were sometimes eaten with bracken fern rhizomes.[49]

Bi was also one of the names for corn (*Zea mays*) when it was introduced.[50]

OCEAN SPRAY	Hanis: nak'áixæɫ, nəq'æixæɫ
Holodiscus discolor	Milluk: nəq'æixæɫ

Plant description: Rosaceae, rose family. Ocean spray, also known as creambush and ironwood, is a shrub that at a glance resembles a small alder when not in bloom, although its leaves are somewhat smaller than alder leaves. In summer this shrub blossoms with sprays of small white flowers that turn brown by late summer or autumn.

Other: Ocean spray wood was carved into tally sticks for scoring in games, like the hoop-and-pole and the hand game.[51] It was also sometimes used to make the playing sticks in the hand game.[52]

Because of its widespread use in making game pieces, ocean spray was one of many spiritual powers that could bring wealth and good luck to people who sought it out.[53]

When it was in full bloom in late July, but before the flowers turned brown, it was time to head to the mountains and hunt bull elk.[54]

Technology: The hard wood of this shrub was used to make a wide variety of tools—digging sticks, stakes, hammers, clubs, arrow points, needles, and combs.[55]

A specialized hammer made of ocean spray wood, called a *mœm-mœm* in Hanis and Milluk, was used to pound maple and cedar bark in preparation for making clothes or baskets. The wood was also dense enough to make a good hunting club.[56]

Many projectile points were made from carefully carved, polished, and heat-hardened points of ocean spray. The points were tapered and sharpened at each end, with one end inserted into the arrow shaft and cemented with a wrapping of eagle feather quills.[57]

A blunt-ended stick of ocean spray was used as a fire drill. To make a fire, a hole was bored in a piece of willow wood, with dry pounded cedar bark added to help start the fire. Then the ocean spray fire drill was rubbed between the hands to create friction.[58]

Sharpened stakes of this wood, 2 to 3 feet long, were placed in the bottom of a camouflaged pit to trap and kill elk.[59]

A section of ocean spray or yew was tied with sinew and pitch to the end of the Douglas-fir shaft of fish spears. This attachment was known in Hanis as *wœ'nəs,* and *háu'yu* in Siuslaw. The sharpened points, of deer or elk bone, were attached to the *wœ'nəs háu'yu.* The hard wood of ocean spray withstood the pressure of using the bone points.[60]

Some stems inserted into pipe bowls for tobacco smoking were made from ocean spray wood, or both bowl and stem were carved from this wood.[61] Other pipe bowls were made of fired clay, schist, or steatite.[62]

OREGON GRAPE
Berberis aquifolium, B. nervosa

Hanis: miyǽcæu aqálqsi, pəgwítłætł aq'álqsi

Plant description: Berberidaceae, barberry family. Oregon grape is also identified in many books as *Mahonia.* These species of Oregon grape are small shrubs, with yellow flowers that develop into tart blue berries. The leaves of both species resemble holly leaves in that they both have sharp points along the leaf margins. The differences are that *B. aquifolium*

(Oregon grape) can be considerably taller than *B. nervosa* (dull Oregon grape)—10 feet versus 2 feet—and dull Oregon grape has three central leaf veins where Oregon grape has one.
Medicine: The roots were used to make a "blood purifier" medicine.[63] This was a common medicinal use by many northwestern tribes.[64]
Other: The Hanis name translates as "rat frightener" or "mouse frightener." If the thorny leaves of this plant were placed in the corners of a house, rats and mice would not enter.[65]

POISON OAK

Toxicodendron diversilobum

Plant description: Anacardiaceae, sumac family. Poison oak is usually a small shrub or vine but sometimes grows as a tall shrub. Leaves are broadly oval, 1 to 4.5 inches long, arranged in threes, and turn bright red in fall. Prefers dry habitats, more common in southern Oregon and California than in northern Oregon.
Other: Frank Drew and Spencer Scott recalled that there had been one small patch of poison oak at Acme (an unincorporated town just east of Florence on the Siuslaw River) that was destroyed when Highway 126 was built. Unfortunately, neither could recall the indigenous names for this plant. Spencer Scott mentioned that he had never seen it at Siletz, but when Indians traveled to the Willamette Valley to pick hops, poison oak was common there and they got "poisoned by it."[66]

RHODODENDRON

Hanis: čʊ́kčʊ́kkwalı

Rhododendron macrophyllum

Plant description: Ericaceae, heath family. Rhododendron is a tall, spreading shrub, common in the understory of dry coniferous forests and on forested sand dunes. Leaves are thick and leathery, 4 to 9 inches long. Blooms in late spring, with large, showy pink flowers.
Technology: Rhododendron roots could be used to carve a shinny ball, which was usually about 2 inches in diameter. In the opinion of Frank

Drew, however, black huckleberry roots made a stronger and heavier ball than those of rhododendron.[67]

ROSE Hanis: muxwtsí'næ
Rosa sp.

Plant description: Rosaceae, rose family. There are several species of wild rose in western Oregon. Two of the more common ones in our region are *R. gymnocarpa*, dwarf rose, and *R. nutkana*, Nootka rose. Both have small pinkish flowers with five petals. The dwarf rose grows up to 5 feet tall; stems are prickly but some young stems are free of thorns. The compound leaves are deciduous, with finely toothed leaflets in groups of five to nine. Nootka rose is similar, but leaves are slightly longer, at 3 inches, and the tips are more rounded.
Other: The wild rose blossom was an inspiration for a basket design.[68]

Numerous tribes in the Pacific Northwest utilized rose hips for food. It is likely that western Oregon peoples did as well.[69]

SALAL Hanis: bá'məs
Gaultheria shallon Milluk: bá'mɪs
Siuslaw: qwan'ní'ɪ (Siuslaw),
kwánni (Lower Umpqua)

Plant description: Ericaceae, heath family. Salal are evergreen bushes that grow along the seashore, on coastal bluffs, and in forests, and they can form dense thickets. Heights range from less than 3 feet up to 15 feet tall. Leaves are alternate and leathery in appearance, up to 4 inches long, with finely toothed edges. Flowers are small, white, and urn shaped. Berries are blue to dark purple, with thick skins.
Food: The berries ripen in late summer and fall. They were eaten fresh. Sometimes the fresh berries were eaten with seal or whale oil.

Many salal berries were dried for winter. One storage method was to pound salal and crab apples into cakes that were dried for winter.[70]

Other: Black dogs were sometimes named "Salal" since the berries are nearly black. When black glass beads were introduced they were also called by the same name as salal.[71]

There was a Hanis myth about how juvenile seagulls came to have dark feathers. A young seagull was once warned not to go to a certain island. The youth went anyway and the people there rubbed salal juice all over him. Since that time juvenile seagulls have had dark feathers.[72]

SALMONBERRY
Rubus spectabilis

Hanis: mí'ya
Milluk: q'æmq
Siuslaw: tł'ʊx, tł ú'ʊx

Plant description: Rosaceae, rose family. Salmonberries are deciduous shrubs, growing up to 12 feet tall, with small thorns on the branches. Leaves have three leaflets with toothed edges. Flowers are bright pink. When ripe, most berries are orange, though some turn dark red.
Food: Salmonberries are the first berries of the year to ripen, around June. They are very watery and do not dry well, so they were eaten only fresh while in season. Salmonberries were also dipped in seal oil, because it helped improve the flavor and because oily foods were supposed to be more nourishing.[73]

The young sprouts were also eaten. Ray Willard (Milluk) recalled that the sprouts gathered were not new growth on mature bushes but were those growing up out of the ground at the base of the salmonberry bushes.[74] This was one of the earliest fresh green foods of spring. In Hanis the sprouts had their own name distinct from the berries, *yúk'wa*. They were peeled and eaten raw or cooked and were often eaten with dried salmon eggs.[75] Frank Drew described the usual treatment of the shoots as follows:

> The women go in April or so, in May perhaps too, or between. The very best shoot in a patch are broken off, and thrown into the large *káwəl* [pack basket] on the back. It's set down in the corner, when returning to the house. They are covered by thorns. A handful of *yúk'wa* sprouts are put on top of the fire, and turned and turned. When about ready, then everybody

> helps himself, takes a sprout, and peels the skin (thorns and all) and eats the inner part of the sprout (with seal grease and salmon eggs as usual). The sprouts are also edible without being cooked. The cooked form has, however, much better flavor when scorched, as described, over the fire.[76]

Other: The calendar was marked by watching when certain plants blossomed and fish runs peaked. At Coos Bay, when the salmonberries began to bloom, flounders were numerous in the lower bay. In the Siuslaw and Coquille Rivers, herrings entered the river when the berries were getting ripe.[77]

When a girl was judged old enough to have some sense and self-discipline, she was told to come along on a salmonberry picking expedition, but she was not allowed to eat a single berry. The girl would return to the village and visit the elders. To each elder, she distributed some of her berries, and each elder would pray over the girl, wishing her a long and healthy life.[78]

It was said that when a person dreamed of salmonberries, he or she would make a new friend, but this friend, like the short-lived berry, would drop quickly. The false friend would last only a short time before making mischief and deceiving one.[79]

Technology: Salmonberry sticks could be used to make the poles in the Coos Bay hoop-and-pole game. The sticks were first peeled and dried. Then they were straightened by working them with hot stones.[80] For a description of the game, known in Hanis as *tlaxaúk'wanawas*, see the entry under red elderberry.

THIMBLEBERRY

Rubus parviflorus

Hanis: tbai
Milluk: tbái

Plant family: Rosaceae, rose family. Thimbleberries are deciduous shrubs up to 9 feet tall and, unlike many other shrubs in the genus *Rubus*, have no thorns. Leaves are up to 12 inches across and have five to seven lobes; short fuzzy hairs on both sides make the leaves soft to the touch. Flowers are small and white.

Thimbleberry, *Rubus parviflorus*.

Food: Thimbleberry sprouts (*mœ'yɪq* in Hanis) were picked at the same time as salmonberry sprouts, in early spring. These sprouts were easier to pick since they are not prickly like salmonberry sprouts. The bark was peeled off and the greens eaten raw or cooked.[81]

The berries were enjoyed fresh, but not preserved for winter. At any one time, few thimbleberries are ripe on any one bush. So women picked unripe thimbleberries and put them in a basket lined with grass. The baskets were kept in the house and covered with leaves. Every morning the basket was opened, and the ripe ones were picked out and eaten while the green ones were sprinkled with a little water.[82]

Other: Baskets for berry picking had some openwork near the top. When a berry basket was full, the berries were covered with thimbleberry leaves or ferns, and sticks were then put through the openwork of the basket, creating a lid that minimized the spilling of berries. It worked well enough that even when a berry basket fell and rolled over, berries were not spilled.[83]

TOBACCO BRUSH Hanis: xwálxwalʊ łǽłəx
Ceanothus velutinus

Plant description: Rhamnaceae, buckthorn family. This shrub has a bewilderingly large number of common names, among them snowbrush ceanothus, tobacco brush, sticky laurel, buck brush, mountain balm, and grease wood. It is a shrub up to 12 feet tall, with somewhat oval leaves with three midveins. Young leaves are somewhat sticky and have a lightly spicy scent. Blooms are clusters of small white flowers. Grows somewhat inland from the coast.

Medicine: The Hanis name means "eye medicine." The roots were gathered, carefully cleaned, and then dried for several days. To treat an irritated eye, a small piece of dried root was soaked in water. The patient was laid down on his or her back, and some of this water dropped in the affected eye. This medicine burned, so only one application of it at a time could be tolerated.[84]

To avoid eye trouble, after killing deer, hunters promptly removed the eyes and ate them.[85]

VINE MAPLE Hanis: hæwǽ'nəs
Acer circinatum Milluk: hǽwæ'nis

Plant description: Aceraceae, maple family. Deciduous shrub or small tree (up to 21 feet tall); leaves have seven to nine lobes and are up to 5 inches across. Leaves turn vivid shades of gold and red in autumn. Seeds are similar to those of bigleaf maple, occurring in winged fruits known as samaras.

Technology: The flexible and tough vine maple was used to make many different household items and tools.

Vine maple was bent into hoops to make the supporting structure for fish traps, a frame for hand drums, and the hoops on baby cradles. Foot-long needles for sewing tule mats were made from vine maple or deer bone.[86]

This flexible wood lent itself to the art of making bows. Frank Drew gave a brief description of the process of making vine maple bows:

Vine maple, *Acer circinatum*.

> The vine maple was cut—a piece as big in thickness as a leg, split into four quarter from the heart. Glass was recently used to scrape a piece down. When in the desired shape it was held on a rack of little sticks high up in a room, and left up there to dry, sort of above the fire. Then the deer sinews was [*sic*] laid on the back and glued on with salmon glue.[87]

Vine maple and hazel were used to construct A-frame smokehouses for drying deer and elk meat.[88]

Vine maple was one of the woods (along with hemlock, cascara, and crab apple) used to make shinny sticks. To shape the end of the stick, it was wrapped with wet grass and steamed in a fire. The steamed wood was limber and could be shaped, although sometimes the stick broke during this process.[89]

Stone pestles have been found in archaeological sites. Many pestles, for pounding berries and roots, were carved from wood, including vine maple.[90]

Jim Buchanan recalled a duck trap made of vine maple, but unfortunately it is not entirely clear how the trap was constructed. He said the bottom sticks of the trap were released by a trigger and caught the bird around the leg or even neck. This type of trap was called *c'hǽi* in Hanis.[91]

WILLOW
Salix sp.

Hanis: k'wǽhæ
Milluk: tłgwɪ, k'wǽhǽi
Siuslaw: ča'átɪ

Plant description: Salicaceae, willow family. Willows, in the genus *Salix*, are deciduous shrubs that have simple, alternate leaves; flowers are catkins. Common species of the Oregon coast are dune willow (*S. hookeriana*), Scouler's willow (*S. scouleriana*), Sitka willow (*S. sitchensis*), and Pacific willow (*S. lasiandra*).

Fiber: Numerous species of willow grow along the Oregon coast, and the bark and roots of many of them could be utilized. Lottie Evanoff said the willow with the "shiny leaf" was used to make baskets, as the switches from a willow with "peach" leaves were not strong enough.[92]

Men peeled willow bark in the summer when out hunting. When the inner bark was separated from the outer bark, the hunters used the inner bark as pack straps and ties for bundles of meat.[93]

The inner bark was also used occasionally for weaving fish trap baskets.[94]

One way of preparing salmon heads utilized willow bark. The inner bark or a spruce root was strung through salmon heads, tied to a stake, and sunk in a creek. After being left for a few days in the creek to age, the heads were retrieved, removed from the string, and boiled into a soup.[95]

Willow switches were occasionally used in basketry, especially for weaving quick pack baskets for meat and salmon.[96]

Other: The season for herring fishing at Coos Bay began when willow just started to bloom.[97]

Pounded-up charcoal from willow was used to make or refresh tattoos. It was believed that willow charcoal made a strong black color and would not fade as quickly.[98]

In Siuslaw, pussy willows were called "little doggies," *šqáxčɪyaič*, based on the Siuslaw word for dog, *šqaxč*.[99]

Technology: In one Coos myth, fire was brought to earth by hiding it in a willow root. To make fire with a drill, the Coos, Lower Umpqua, and Siuslaw all used the same method. A hole was bored in a piece of willow wood. A drill of ocean spray, or perhaps willow, was placed on the willow wood and rubbed between the hands. Pounded dried cedar bark or moss was placed around it to catch the fire.[100]

7 Forbs

BEARGRASS
Xerophyllum tenax

Hanis: qǽ'ætł
Milluk: qætł'ən
Siuslaw: c'xwis (Siuslaw),
pax (Lower Umpqua)

Plant description: Melanthiaceae, bunchflower family; formerly classified in the Liliaceae, lily family, and still identified as such in many general plant guidebooks. Its leaves resemble a tuft of tough grass. Every few years, it blooms, sending up a stalk 3 to 4 feet tall, crowned with a dense cluster of tiny white flowers. It prefers open meadows in the Coast Range and Cascades, although in the past it was also found at lower altitudes near the coast, where it is rarely seen today.

Fiber: The leaves were harvested in summer, dried and bleached white in the sun, and used as decorative white overlay in basket designs. They were also sometimes used to add designs to the edges of leggings.[1]

This plant was collected in the Coast Range and historically near the coast. The Siuslaw people used to harvest much of their beargrass at a place 3 miles west of Elk Prairie, between Noti and Walton.[2]

Other: In Beverly Ward's memoir, she learned from her husband's Lower Coquille grandmother, Susan Ned, that Indians cleaned themselves with sand and soap made from roots of "squaw grass" (which is probably a reference to beargrass).[3]

BRODIAEA, HARVEST
Brodiaea terrestris ssp. *terrestris*, *B. coronaria*, *Dichelostemma congestum*

Hanis: wíllɪts', wúlæc'
Milluk: wɪllɪc'
Siuslaw: k'wʊsk'w (small camas)

Plant description: Asparagaceae, asparagus family. Formerly classified in Liliaceae, lily family. Slender basal leaves; purple liliaceous flowers in open umbels, with prominent midribs down the center of each petal, blooming generally May–July. *Brodiaea coronaria* (harvest brodiaea) stems can be up to 14 inches tall, whereas *B. terrestris* ssp. *terrestris* (dwarf brodiaea) stems are 2 inches, or even smaller. *B. coronaria* is found from British Columbia to California, while dwarf brodiaea grows from Coos County to southern California.

Food: The corms of brodiaea were a traditional food of many tribes in the Pacific Northwest and California. The Coos, Lower Umpqua, and Siuslaw harvested the corms in late spring or early summer. The corms could be dried and preserved for winter stores, and cooked in earth ovens or boiled. Although the two species of brodiaea often grew in or

Blue dicks, a type of brodiaea harvest lily species, *Dichelostemma congestum*.

near the same meadows as camas and were gathered at the same time, the corms were stored separately from camas.[4] Frank Drew said of brodiaea, "*wulœc*' is a root similar to camas, but very tiny, round . . . it is better eating [than camas]."[5]

Dwarf brodiaea, *B. terrestris* ssp. *terrestris*, grew at Gregory Point, Cape Arago, and along the coastal plains with camas down to the Coquille River. A patch of two closely related species, *B. coronaria* and *Dichelostemma congestum* (formerly *Brodiaea congesta*), grew at a place near the junction of the Siuslaw River and the North Fork Siuslaw, and this place was named for these plants, *k'wʊ́sk'wyamus*.[6] A century ago, *B. coronaria* was also observed in sand dune meadows.[7]

In English, brodiaea was always described as a variety of small camas by many western Oregon Indians, based on the similar appearance of the plants.[8]

BULRUSH, PANICLED Hanis: wǽ'ætɬ'
Scirpus microcarpus Milluk: wæ'ǽtɬ'

Plant description: Cyperaceae, sedge family. Tall, grasslike, triangular in cross section, edges of leaves sharp. Widespread in wet habitats along streams and in wetlands.
Other: Thatched roofs of this bulrush, colloquially known as cut grass, were used on temporary shelters, houses, and storage sheds. The wider root ends of the bulrush leaves were laid between the poles to create the thatching. These were then laid on a frame of support posts and rafters. Bulrush could also be used to build the walls of these structures, along with thatching of sword fern.[9] According to Lottie Evanoff, bulrush thatching never leaked.[10]

BUR-REED Hanis: xúnɪs
Sparganium eurycarpum, Milluk: xúnəs
S. emersum

Plant description: Typhaceae, cattail family. Perennial; *S. emersum* grows up to 20 inches tall, *S. eurycarpum* up to 3 feet tall; both are emergent at the edges of ponds, lakes, and sloughs. Spherical, burred flower heads bloom in mid to late summer.

Other: Bur-reed is a marsh plant that resembles cattail, until it develops its seed head. In a myth told among the Coos, Lower Umpqua, and Siuslaw, Beaver passed off bur-reed and willow limbs as salmon and trout for his wives. His disappointed spouses soon abandoned Beaver for a wealthier and more competent man.[11]

CAMAS	Hanis: qǽ'mæ
Camassia quamash,	Milluk: qæm
C. leichtlinii	Siuslaw: aučísi

Plant description: Asparagaceae, asparagus family. Until recently, camas was classified in the lily family, Liliaceae. These two species of camas are very similar. They are perennials that grow from bulbs; flowers have six petals, usually blue purple, rarely white; stamens yellow. Grows in moist meadows.

Food: Camas bulbs were one of the most important root foods for Native Americans in the Pacific Northwest. *Camassia quamash* seems to be more common at the coast, with *C. leichtlinii* found inland mixed with *C. quamash.*

Two harvest times for camas are mentioned. One is early spring when the plants are in flower, and the other is in summer after the flowers have died back.[12] Indeed, people may have thought it was important to let the plants seed before harvest. Coquelle Thompson (Upper Coquille) mentioned, "When woman dug camas, they just let seeds go."[13]

At harvest time, a large group of people would set out for the camas prairie to dig the bulbs and build earth ovens. Frank Drew gave a description of women harvesting camas:

> The arrowwood [*Holodiscus discolor*] shovel is used to dig them slow work it is. A days work is 40–70 pounds. They are packed in the *kawəl* [burden basket]. . . . Camas must be "barbecued" roasted under earth in an earth oven. When taken out of the steaming heat of the ground oven they are more firm and have changed to a sort of yellow color. Then they can be eaten or

Women gathering camas near Elkton, Oregon. Sketch by Captain Albert Lyman, 1850. Image courtesy of Douglas County Museum.

> kept indefinitely in a warm place in the house. Sometimes a handful will be taken, and, being soft, can be mashed in the hand into a ball, and eaten like a potato, with seal grease. They are kept [through] the winter in the storage basket made of hazels or young willows.[14]

The earth ovens were lined with rocks, which were first heated by building a large fire. After the rocks were hot and the fire ashes swept away, layers of bark, sweetgrass, fern leaves, and herbs were layered into the oven. Then the camas bulbs were put in, and more layers of grasses and leaves were placed on top of them. This was then topped with sand, and a fire was built on top of the oven. The camas bulbs were cooked for a full day. They could be eaten promptly after removal from the oven, or they could be peeled and pressed into "cakes" (known as *hamai* or *tłax* in Hanis) for storage in grass-lined baskets through winter.[15] These pressed cakes of camas were described as having a consistency like that of cheese.[16]

Since camas was an important staple, many gathering sites were recalled. There were camas meadows around Whiskey Run and Cut Creeks, where Coos Bay and Lower Coquille Indians met at an annual rendezvous. Other patches in the Coos Bay area were at Cape Arago and

Tarheel, on the lower bay.[17] It was said that not long ago, the outlet of Coos Bay was at Jarvis landing, and there was a camas patch where the current outlet is now.[18]

An important regional camas prairie was at Camas Valley, near the headwaters of the Middle Fork of the Coquille River. According to Coquelle Thompson, an Upper Coquille man whose paternal grandfather was Upper Umpqua, Camas Valley was originally inhabited by a band of Upper Umpqua people known as the *sɛtł'umɛ́dunnɛ* (a southwestern Oregon Athabaskan name meaning something like "people of the damp oak prairie").[19] Lottie Evanoff said that her father, Chief Doloos Jackson, described it to her as a place "blue with flowers and only trees there were oaks."[20] Indians from all around the region probably gathered here to dig camas, trade, and socialize.

Lower Umpqua people probably got most of their camas from prairies not far upriver from *C' álila*, the uppermost Lower Umpqua village, located a short distance upriver from Wells Creek. There was also a camas prairie at Loon Lake, probably in Ash Valley to the north of the lake. There are stories that grizzly bears gathered camas at Loon Lake too, and people had to be careful to stay away from camas gathered by the grizzlies, or be mauled by the bears.[21]

There is little mention of camas in Siuslaw country, and they probably got much of it from the Kalapuyans in the Willamette Valley. The Siuslaw word for Kalapuyans—*áuča*—is based on their word for camas, *áučisɪ*.[22]

There is also a camas patch on the east side of Cape Perpetua. This meadow originally belonged to the Alsea people at the nearby village of *Yáxaikʸ*, but during the years of the Alsea subagency the Coos and Lower Umpqua people harvested from this meadow.[23]

Indeed, the camas prairie was so important that its creation was told in an Alsea myth (probably well known to the Siuslaw as well) that as *S'úku*, the world transformer, traveled along the coast bringing salmon to all the rivers, he stopped briefly at Perpetua and Yachats:

> Then he went on and came to where a mountain was touching the edge of the water [Cape Perpetua]. So he climbed up (there). And after he came to the top he looked around and perceived a pretty valley. So he began to climb down. And after he came

> down he went on. But he did not go (very long) and said "How would it be if I should climb up for a little while and look at the place once more?" So then after he came to the top he went to where that pretty place (was). . . . "I am going to break wind right here, so that the place may have camas." Then after he finished (doing this) he went down again. . . . Then as soon as he came to that prairie [Yachats] he walked around in different directions and began to break wind all over the place. That is the reason why the [Yachats] prairie has camas all over, because he did so (at that time).[24]

The connection between *S'úku* "breaking wind" and camas is probably more than just injecting some Trickster humor into the story, as eating camas can cause intestinal gas.[25]

Other: According to Billy Metcalf, Chetco/Joshua, the people of the lower Rogue and Chetco Rivers rubbed raw camas bulbs on baskets to make them watertight.[26]

CAMAS, DEATH

Toxicoscordion fremontii

Plant description: Melanthiaceae, bunchflower family (formerly classified in the lily family). Leaves grasslike, up to 1 foot long; flowers white, in clusters, bloom in May or June. When plant is not in bloom, its leaves and bulbs resemble those of common camas.

Medicine: Death camas grows in the same habitat as camas. Care was taken to keep camas patches cleared of death camas, because both leaves and bulbs of the latter contain toxic alkaloids and can cause death if ingested. The bulbs and leaves closely resemble those of the two edible species of camas.

Beverly Ward, writing about the Lower Coquille people, mentioned that the Indians were aware of the toxic qualities of death camas but used it as medicine. Unfortunately, she did not specify what kind of medicine.[27] The Chehalis and Squaxin peoples of western Washington applied a poultice of the bulbs to sprains, bruises, boils, rheumatism, and general pains. Some even ingested a little of it as a violent emetic.

Perhaps these uses were known to Oregon natives. However, because of its toxicity, it is best to avoid ingesting any part of this plant, under any circumstances.[28]

CATTAIL
Typha latifolia

Hanis: łhwai (leaves), łə́'læm
Milluk: łhwai (leaves), łəlæm
Siuslaw: təmk'ʊ́lla

Plant description: Typhaceae, cattail family. Cattail grows 3 to 9 feet tall in shallow fresh or brackish water. It develops a distinctive brown spike at the top of its stalks.

Fiber: Cattails provided an important and versatile fiber for basketry. The leaves were split and the fibers used as warp and weft for many varieties of twined baskets, from large pack baskets to women's hats.[29]

The basket caps of Coos, Lower Umpqua, and Siuslaw women were quite different from those made by their neighbors to the south. Cattail leaves were used for warp and weft, sometimes including *Juncus* rush for the warp too, and for decorative overlay many other materials such as beargrass and eelgrass were used. The people of the southern Oregon coast and northwestern California used willow or hazel sticks for the warp, conifer or willow root for the weft, and materials such as beargrass, maidenhair fern, and chain fern for overlay. Women often traded caps and were familiar with both styles.

The leaves could also be rolled into twine and used to sew tule mats, weave dip nets, and make rope.[30]

Cattail leaves were gathered in summer, when they were described as being "strong" enough to be cured in the sun. From there, dried strips of cattail leaves could be stored until needed for basketry or twine.[31]

Cattail stalks (and perhaps the leaves as well) were pounded until they were fine and soft, and used as diapers.[32]

Cattails also made raincoats. The long leaves were sewed together, tied around the shoulders, and worn by men and women when they worked in rainy weather. It was said that a person would not get wet, the leaves shed rain so well.[33]

Food: The roots of cattail were gathered in spring. They were good to eat either raw or cooked.[34]

Medicine: Mashed-up cattail was warmed and rubbed on the noses of infants to prevent them from having flat noses. The infants' arms and legs were then rubbed and massaged to help them grow straight. After the massage, the infants were bathed.[35]

Susan Ned, a Lower Coquille woman, mixed cattail tops with lard to treat burns.[36]

Other: Although cattails are common in marshy areas, they were perhaps especially abundant in Pony Slough, on Coos Bay. There was a site on Pony Slough called *łhwáhich*, meaning "cattail place."[37]

CLOVER, SPRINGBANK

Trifolium wormskioldii

Hanis: yǽ'æt
Milluk: yǽ'æt
Siuslaw: q'win

Plant description: Fabaceae, pea family. Perennial clover, leaves up to 1 inch long, pointed, with finely toothed margins. Flowers pinkish, often with whitish tips, in a dense head up to 1 inch across. Grows in moist to wet open meadows.

Food: The roots of this clover, usually gathered in autumn, were a popular staple of many coastal peoples of the Pacific Northwest.[38] The roots were boiled or baked in earth ovens before they were eaten. They could also be sun dried and preserved for winter.[39]

Frank Drew had a good description of the gathering and cooking of springbank clover:

> The women use the arrowwood [*Holodiscus discolor*] shovel. It is taken out; they are thick and plentiful. It takes a lot of time to get enough because they are each so small. The leaves are thrown away, the roots are put loose into the basket, not bundled. They have to be cooked in an underground fire covered with grass and earth overnight. Sometimes 3–4–5 women each share in the oven, with their diggings. When uncovered and laid out to cool, while still hot but cooling they wriggle and squirm. Then they may be eaten cold, with seal grease, in any of the next few days.[40]

Spencer Scott enjoyed this food and said the roots were "whitish, sweet to the taste." He recalled that the Alsea people also ate the roots and called them a name similar to the Siuslaw term *q'win*.[41]

Other: Dried stems of wild clover and a wild lettuce were gathered by children and threaded for necklaces as imitation dentalium. Kinnikinnick berries were used as beads. Annie Peterson called it "playing Coyote," as sometimes the Trickster would dress up in these stems, berries, and red currant blossoms as imitation finery.[42]

COW PARSNIP
Heracleum maximum

Hanis: qá'yaq', káyaq'a
Milluk: qá'yaq'
Siuslaw: q'ápɪch

Plant description: This species is also known as *Heracleum lanatum*, in the parsley family, Apiaceae. Large perennial that can grow almost 10 feet tall. Leaves can grow to be more than 12 inches across, divided into three lobes. Flowers white, in a flat-topped umbel. Mature plant has a strong, pungent smell.

Food: Cow parsnip sprouts were widely eaten in the Pacific Northwest. In early spring, when the upper stalk was tender, it was broken off, the outer skin peeled away, and the stalk eaten. Often it was dipped in seal grease before it was consumed. The outer skin of the plant had to be removed, as it is bitter and causes a burning sensation in the mouth.[43]

People were aware of the toxicity of western water hemlock, *Cicuta douglasii*, which resembles cow parsnip. When they gathered cow parsnip, they took care to avoid accidentally harvesting its poisonous cousin. Spencer Scott commented that water hemlock was darker in color than cow parsnip.[44]

The Rogue River and Chetco peoples also used cow parsnip as a medicine. The roots were crushed and put on hot rocks and the steam inhaled to treat head colds. The roots were also used as a liniment for aching joints.[45]

CUCUMBER, WILD Hanis: bíga
Marah oregana Milluk: bíga

Plant description: Cucurbitaceae, gourd family. Trailing, climbing perennial herb, with curly tendrils. Leaves broad, with five to seven lobes. Flowers white and star shaped. Fruit round, green, covered in distinctive soft "spikes."
Medicine: The root was boiled and the water bathed in for rheumatism.[46]
Other: If a darkened seed of the wild cucumber was found, it was taken home to bring good luck. It was believed to be the eye of a giant mythological bird, known in Hanis as *yágals*.[47]

DOCK Hanis: mǽhɪ
Rumex sp. Milluk: mǽhɪ

Plant description: Polygonaceae, buckwheat family. Docks are "weedy" herbs, often growing in disturbed soils. Most species have long, lanceolate leaves, concentrated mostly along the bottom portion of the stem. Flowers are small, in dull colors of reds, greens, and browns and packed in tight clusters. Indigenous species of dock include golden dock (*Rumex maritimus*), willow-leaved dock (*R. salicifolius*), and western dock (*R. aquaticus*). Other common docks introduced from Eurasia include sheep sorrel (*R. acetosella*) and yellow dock (*R. crispus*).
Medicine: Annie Miner Peterson said the roots of a plant she called yellow dock were pounded to make a poultice to treat aching bones.[48] It is not clear which species of dock she was referring to. *Rumex crispus* is usually known as "yellow dock" (also "curly dock") but is widely thought to be introduced from Eurasia. Nevertheless, this species was adopted by many North American tribes for food and medicinal use, as was the indigenous western dock (*R. aquaticus*).[49]
Other: Susan Ned, a Lower Coquille woman, said that most Indians smoked kinnikinnick and tobacco, but a few smoked the leaves of "plantain and broad leaf dock."[50] Usually the common name "broad-leaf dock" refers to *Rumex obtusifolius*, a species introduced from Europe.

Carol Batdorf noted that Coast Salish people used to smoke "dried dock leaves" of *R. crispus* and *R. occidentalis*.[51]

EELGRASS
Zostera marina, Phyllospadix scouleri

Hanis: qwaq'w, łqálqas
Milluk: ləmə́n

Plant description: Zosteraceae, eelgrass family. *Phyllospadix scouleri* (surfgrass) and common eelgrass (*Zostera marina*) closely resemble one another. Both grow in saltwater from the low tide zone to the subtidal zone, have bright green grasslike leaves, and can form substantial colonies. However, they grow in rather different habitats—*Zostera marina* is found in sandy to muddy habitats in estuaries, while *Phyllospadix scouleri* grows in rocky substrates. *Zostera marina* leaves are about 1 inch wide and can grow up to 12 feet long, although they are usually much shorter. They leaves of *Phyllospadix scouleri* are narrower (typically ¼ to ½ inch) and in most populations grow to lengths of 3 feet or less.

Populations of *Phyllospadix scouleri* are reported for the rocky shores at Cape Arago, while populations of *Zostera marina* are found in the estuaries of Coos Bay and the Umpqua and Siuslaw Rivers.

Fiber: Eelgrass was gathered and blackened while it was still fresh by holding it in the smoke of a fire built with pitchy wood. The blackened eelgrass was used as a decorative overlay in baskets by peoples all along the Oregon coast as far south as Coos Bay. The Athabaskan peoples to the south used maidenhair fern for black overlay instead.[52]

Siuslaw elder Dorothy (Barrett) Kneaper recalled that her grandmother would pick up slippery grasses that washed up along the shore of the Siuslaw River to make tiny baskets for her grandchildren. This was probably *Zostera marina*.[53]

FALSE LILY OF THE VALLEY
Maianthemum dilatatum

Hanis: c'ə́nnau wəxǽya

Plant description: Asparagaceae, asparagus family; formerly Liliaceae, lily family. This plant is also known as May lily, deerberry, and

False lily of the valley, *Maianthemum dilatatum*.

bead-ruby. Individual plants grow no more than 12 inches tall and can form extensive carpets in the forest understory. Each plant has two to three heart-shaped shiny green leaves. Blooms in early summer with clusters of small white flowers. Berries are small and round, covered with reddish spots when ripe.

Other: The Hanis name translates as "thunder's dried things," but unfortunately the story behind the name was forgotten.[54]

In Coos cosmology, thunderbirds were chiefs of the sea and its denizens. Interestingly, the Quileute people of northwestern Washington thought that the berries of this plant were oily, and one of their names for the plant associated it with whale oil.

The berries are edible, and some northwestern peoples ate them, but sparingly, and nowhere do they seem to have been relished.[55]

FIREWEED Hanis: łnakátı
Chamerion angustifolium Milluk: łnakátı

Plant description: Onagraceae, evening primrose family. Fireweed is also known as willow herb and was formerly classified in the genus *Epilobium*. A perennial, it can grow up to 7 feet tall, with lanceolate leaves up to 5 inches long. The leaf veins are distinctive. They do not

Fireweed, *Chamerion angustifolium.*

extend to the edge of the leaf but end at another leaf vein that roughly parallels the edge of the leaf. Blooms in summer with pinkish-red flowers. Fireweed grows on lands cleared by fire or logging and along roadsides.

Fiber: This plant was harvested in early summer, and the outer stem fiber was peeled, dried, and made into a fine twine. The twine was then used to make a dress. It took so much fiber to make these dresses that only women in wealthy families wore them, and they wore them only for special occasions such as dances, feasts, or weddings. Poor women gathered the fireweed and sold it to wealthy women. The thread used for twining and sewing the dresses came from an unidentified plant that Annie Miner Peterson described as flowerless and only 4 to 5 inches tall. The dresses were then trimmed with the valuable scalps of the red-headed woodpecker.[56]

Many Northwest coast people used fireweed to make cordage, although it was often considered inferior to other sources of fiber. Usually, fireweed stems were gathered after flowering. However, the Coos, Lower Umpqua, and Siuslaw gathered the plants before they flowered for processing into fiber—sometime in June or July. Using such “green” fireweed for cordage, while unusual, is not wholly unknown elsewhere—the Gitxsan people of British Columbia also did this and used it for cordage and nets.[57]

Medicine: The roots of fireweed were used to make a poultice for lower back pain and swollen muscles and sprains. The roots were cleaned, stripped of bark, pounded, and then wrapped in buckskin or, in more recent times, fabric. This poultice was left overnight on the affected area, and in the morning the swelling and soreness were supposed to be gone.

It was believed that putting the root in one's mouth would cause a bad tooth to completely decay.[58]

GINGER, WILD

Asarum caudatum

Hanis: ǽqæu k'whánas

Plant description: Aristolochiaceae, birthwort family. An evergreen forb that grows in the deep leaf litter of damp coastal forests. The leaves are heart shaped, with a velvety texture. The stems sprawl along the forest floor. The brownish-red flowers are somewhat bell shaped, with pointed tips. The scaly rootstocks smell like ginger.

Medicinal: A poultice of this plant (parts unspecified) was used to treat backaches.[59]

Wild ginger, *Asarum caudatum*.

Other: The Hanis name of this plant means "dead person's ear." If a gun was failing to get game as well as it used to, many wild ginger leaves were gathered and boiled. The inside of the gun was washed out with a ramrod soaked in this wild ginger tea, followed by a cleaning with elk tallow. It was believed that this doctoring would make the gun more effective. Presumably, this practice was done for bows before the introduction of guns.[60]

IRIS Hanis: tæmǽlæ
Iris douglasiana

Plant description: Iridaceae, iris family. This iris grows from southern coastal Coos County south to southern California. Leaves are at least 1 inch wide and up to 18 inches tall. Stems are slightly taller and bear one to three flowers, which are usually purple but are sometimes lavender or white.

The tough-leaved iris, *Iris tenax*, is similar but has narrower leaves. It grows from southwestern Washington to northwestern California.

Technology: The leaves of this plant were gathered, dried, and then pounded with a wooden pestle on a board to carefully separate the fibers. These fibers were twisted into cordage that was tough and strong. It was the preferred cordage for fishing line, dip nets, and snares, as it was said to never break. The plant grew mostly south of Coos Bay and in some of the inland valleys, and most of the leaves were obtained in trade with the Upper Coquille and Rogue River tribes.[61]

The Hanis word for this iris may have been borrowed from Athabaskan-speaking tribes to the south along with the plant itself. In 1885, James Owen Dorsey recorded the word *te-me-le* in Naltunnetunne (an Athabaskan dialect from the southern Oregon coast) for a "grass" that was used to make nets and snares—almost certainly it was this iris, the leaves of which do resemble those of grasses.[62]

David Douglas observed that the Upper Umpqua made a strong snare from the leaves of a "small species of Iris, found abundantly in the low moist rich grounds." This species may have been *Iris tenax*.[63]

LILY, TIGER
Lilium columbianum

Plant description: Liliaceae, lily family. Perennial lily, with narrow, lanceolate leaves up to 4 inches long, and yellow-orange flowers with dark red or purplish spots near the center.

Food: Tribal elder Dennis Rankin recalls seeing tiger lilies growing on his grandmother Rose McArthur's family allotment on the lower Umpqua River. She told him that the Lower Umpqua people gathered the bulbs and boiled them.[64]

The practice of gathering tiger lily bulbs for food was very widespread in the Pacific Northwest. Some authors state that the bulbs are bitter and peppery and so were used more as a spice or condiment than a staple food. In British Columbia, the bulbs were gathered at many different times of the year—in spring before flowering, in summer during flowering, or in fall after the flowers had died back.[65]

LUPINE, SEASHORE
Lupinus littoralis

Hanis: ha'údıt
Milluk: ha'wádıt
Siuslaw: k'á'asa (Lower Umpqua), q'axc' (Siuslaw)

Plant description: Fabaceae, pea family. This plant was also known as Chinook licorice. Seashore lupine is a perennial that grows on beaches and sand dunes. The stems and palmately compound leaves are covered with small hairs that give them a silvery or gray color. Flowers are purple and pealike.

Food: After the leaves died back in fall, the roots were gathered and roasted in earth ovens. This plant was mentioned by many and seems to have been a staple food. It was said to taste like sweet potato, although in English some Indians called it "wild carrot" (a confusing term since

Seashore lupine, *Lupinus littoralis*.

"carrot" usually refers to a member of a different family, the Apiaceae). This plant grows profusely in the sand dunes and was often gathered there.[66]

This plant was eaten by many coastal tribes. Lewis and Clark ate it during their stay with the Chinooks. The name "Chinook licorice" came about because some whites thought the roots had a licorice-like taste to them. The roots of all lupines, including this one, contain toxic alkaloids, but cooking the roots seems to make them harmless. The Kwakwaka'wakw people of coastal British Columbia believed that if people ate the raw roots, they would act drunk and sleepy.[67]

MINER'S LETTUCE Hanis: wál'walʊ łǽłəx

Claytonia perfoliata,
C. rubra ssp. *depressa*

Plant description: Family Montiaceae (formerly Portulacaceae, purslane family). Red-stem miner's lettuce, *Claytonia rubra* ssp. *depressa*, was formerly classified as a subspecies of *C. perfoliata*. Miner's lettuce (also known as winter purslane or Indian lettuce) is a small fleshy annual, growing no more than about 1 foot tall. Its leaves are somewhat unique—they are opposite but partially fused, so it appears that the plant's stem grows up through the middle of the leaf. Flowers are small and white. In red-stem miner's lettuce, the undersides of the leaves are a dark reddish color.

Food: Grace Brainard, married to a Milluk Coos man, recalled picking and eating a plant she called "deer lettuce," which, from her description (common coastal plant with small, heart-shaped leaves and small white flowers, good in salads), sounds like miner's lettuce.[68]

Medicine: The literal translation of the Hanis name is "knife's medicine." The stem and leaves were placed over a cut after the blood had been washed away.[69]

One treatment for rheumatism was to make shallow cuts (with flint or glass) around the affected joint. Then the blood was scraped away with a stick and the cuts covered in a poultice of miner's lettuce. The bloodied stick was then thrown into a river. Bruises were treated in a similar manner.[70]

Grace Brainard learned from her husband's Milluk grandmother, Frances Elliott, that a tea made by boiling wild lettuce was good for treating eye infections.[71]

Annie Miner Peterson mentioned a "wild lettuce" that was used as a poultice to treat swellings and was also drunk as a tea. It was probably also *Claytonia perfoliata*, or possibly a closely related species, *C. sibirica*.[72]

NETTLE
Urtica dioica

Hanis: walláq'as
Milluk: wálaq'as

Plant description: Urticaceae, nettle family. This herb grows in moist, nitrogen-rich soils. It can grow as tall as 6 feet but more typically reaches heights of 3 feet. Young plants often have reddish-tinged leaves. Its leaves are opposite, with coarsely toothed margins. The undersides of the leaves and stems have small hollow hairs that contain an acid. Touching them causes a mild and uncomfortable burning sensation.

Medicine: The pounded roots were placed on aching joints to treat rheumatism.[73]

Other: I have found no references to Coos, Lower Umpqua, and Siuslaw peoples eating nettle greens in precontact times, but some do today. Grace Brainard recalls gathering nettles in spring with her Milluk husband, Emil Brainard, to eat. She said that near the nettle patches were plants with furry leaves that when rubbed on the skin took away the stinging sensation caused by touching the nettles.[74]

Some tribal members today utilize dried nettle leaves for tea.

ONION, WILD

Allium cernuum

Plant description: Liliaceae, lily family. Wild onions are perennials, growing up from clusters of bulbs that have a strong onion smell. Leaves are about 2 feet tall and resemble grass. The pink or light purple flowers are bell shaped and occur in drooping clusters.

Food: Wild onion bulbs could be eaten raw or cooked. Formerly, many wild onions grew on a prairie a short distance east of Bandon. Unfortunately, no one could recall the indigenous names for the wild onion.[75]

PIPSISSEWA

Chimaphila umbellata

Plant description: Ericaceae, heath family. Pipsissewa is also known as prince's pine (occasionally mistakenly written as princess pine). It is a small, slightly woody herb that grows up to 15 inches tall. Leaves are up to 3 inches long and shiny, with lightly toothed edges. Flowers are small, pale pink to rose, waxy, in loose clusters of three to fifteen. Grows in coniferous forests in well-drained soils and rotten logs at low to moderate elevations in the mountains.

Food: The dried leaves of this plant were used to make a tea.[76]

Pitcher plant, *Darlingtonia californica*.

PITCHER PLANT (COBRA LILY)
Darlingtonia californica

Hanis: qwsáálʊq'w, kwɪssalʊk
Milluk: qwsáálʊq'w

Plant description: Sarraceniaceae, pitcher plant family. Pitcher plants, also known as cobra lilies, are carnivorous plants that grow in sphagnum bogs and serpentine fens in western Oregon and northern California. Pitcher plants can grow up to 3 feet tall, with leaves that form a tube resembling a cobra's head. The tube is mottled with translucent "windows" that help confuse insects that have been lured inside.

Other: The plant was picked and used as a cup when a person was out hunting and had nothing else to use for this purpose.[77]

Many Coos people in the late nineteenth and early twentieth centuries harvested cranberries at McFarland's cranberry bog, located north of Coos Bay near Hauser. There were many pitcher plants near McFarland's bog, and the Hanis people nicknamed his place *kwíssalʊkwa*.[78]

POND LILY Hanis: qá'waltł
Nuphar polysepala

Plant description: Nymphaeaceae, water-lily family. Yellow pond lily is common in shallow ponds and lake margins. It is an emergent aquatic plant, with large leaves (up to 7 inches across) and round yellow flowers.
Other: No cultural use was mentioned by the Coos, Lower Umpqua, and Siuslaw.[79] However, in one version of the story of Beaver and Muskrat, Beaver gathered a war party together to attack a rich man and win back his wives, who had deserted him. Beaver went ahead carrying a large knife while his allies waited in bushes to ambush the wealthy man. He told them that if they smelled pond lilies, he had been killed. Beaver was killed, and his allies smelled pond lilies, so they all went home. Meanwhile, the young man put the knife in Beaver's backside and threw him in the lake. Since then, all beavers have large tails.[80]

Some tribes ate the parched seeds of this plant. The roots are said to be edible as well, but bitter. Few tribes used the roots for food, but some did use them for medicine. Dr. Walton Haydon, a medical doctor and amateur botanist who lived in Coos Bay about a century ago after working many years for the Hudson's Bay Company, noted that the roots were edible and "one of the common foods of the indians [*sic*]." He did not make it clear whether he was referring to his earlier experience among the Cree and other natives during his years with Hudson's Bay, or whether he referred to the Coos and their neighbors.[81]

RICE ROOT Hanis: čahwənnia
Fritillaria affinis

Plant description: Liliaceae, lily family. Rice root is also known as chocolate lily after its brownish, yellow-speckled flowers. Stems grow to 30 inches tall; leaves are alternate, lance shaped, and 2 to 7 inches long. Habitats are meadows, bluffs, and open woods.
Food: The bulblets on the roots of this plant resemble rice. The roots were dried and stored for winter.[82]

Many tribes in British Columbia also cooked and ate the roots of this lily.[83] Coquelle Thompson mentioned a plant people gathered near Corvallis that may have been rice root, or a near relative. He said this plant had "edible roots, grow 2 feet height, brown-red flowers, bring back with camas. Roots drab color, grow in fields at Corvallis, none at Siletz."[84]

The roots have often been described as having a slightly bitter taste. Leslie Haskin, who wrote one of the first botany books in the Northwest for a popular audience, cooked and ate a batch of the bulblets. He found them to be "tender and delicate, and except for a slightly bitter taste they could scarcely be told from genuine rice. With cream, or butter and sugar, they are very fine."[85]

RUSH
Juncus sp.

Hanis: bábəs, dínı
Milluk: dínhı

Plant description: Juncaceae, rush family. *Juncus* rushes resemble grasses, and they grow in clumps or extensive colonies in sand dunes and damp fields. Locally they were sometimes called "wire grass" or tussock. Their flowers are small and brown, growing near but not at the top of the stems.
Fiber: Rushes were used to make tump lines for pack baskets, rope for berry baskets, and carrying straps for baby cradles. They were also sometimes used along with cattail leaves in weaving women's hats, and also in berry baskets with beargrass as an overlay.[86]

This rush was often found in sand dunes and marshy places.[87]

The word *bábəs* was borrowed from Upper Coquille Athabaskan; these people also used it in basketry.[88]

SANDMAT
Cardionema ramosissimum

Milluk: cgánatł'

Plant description: Caryophyllaceae, pink family. Sandmat is a ground-hugging plant, with numerous branches growing up from a taproot.

Grows near beaches and in the dunes. Leaves are tipped with sharp thorns. Flowers are small and white.

Other: In Annie Miner Peterson's telling of the cycle of five Tricksters, the fifth Trickster created this prickly plant. He has married Moon, and her parents keep trying to trick and kill him, but he always outwits them. Finally, his mother-in-law tries to drop fine "fir-bark-stickers" down on him:[89]

> Now inside fine-fir-bark-stickers she poured down on him.
> Now he blew it up, the young man.
> Now she got her vagina full of it,
> when the young man blew up those bark-stickers.
>
> Now the old woman fell down, she fell belly up.
> She kicked [in her pain]
>
> The young man said [to the Moon girl]
> Go wash your mother with water.
>
> Then the girl washed her mother with water.
> Now the old woman was all right again.
>
> Now the young man said,
> "Hereafter, they [stickers] will be on the sand.
> They will see them there [on sand beaches].
> They will grow [on the sand] there.
>
> So it is indeed.
> That thing grows on sand now,
> That tree-sticker [*cgánatł'*]

Melville Jacobs thought that this plant was a "cactus or thistle of some sort." There are no species of cactus native to the Oregon coast. The likeliest plant that fits this description is sandmat, which is extremely prickly and can be found in dunes and along beach margins.[90]

SKUNK CABBAGE
Lysichiton americanus

Hanis: yayáx, yáiyax
Milluk: kímætł'
Siuslaw: c'yá'nax, c'yáánx

Plant description: Araceae, arum family. There are no other plants in western Oregon quite like skunk cabbage. It is a wetland plant, growing in muddy, mucky soils in swamps. Its glossy leaves emerge early in spring and can reach a height of 4 feet. The flowers are greenish yellow and are on a clublike spadix that emerges from the center of the leaves. The spadix is surrounded by a bright yellow spathe (modified leaf). Skunk cabbage blooms produce a distinctive smell that many people find unpleasant.

Food: The roots were roasted in hot ashes or earth ovens and often eaten with seal oil. Roasting in hot ashes took three to four hours. Prolonged cooking was necessary, as the raw roots are full of calcium oxalate, which causes a burning sensation in the mouth. The roots were fully cooked when they smelled sweet. Properly cooked roots tasted like cabbage. The preferred roots were young ones, so the best time to gather them was in early summer.[91]

Lottie Evanoff noted that bears liked to eat skunk cabbage, and "everything bear eats is good eating." She thought skunk cabbage roots were good and found it curious that white people did not eat them.[92]

Medicine: Frances (Talbot) Elliott, the author's great-grandmother, used to make a cold medicine by grinding up raw skunk cabbage root and mixing it with honey.[93] Beverly Ward recalled that her husband's Lower Coquille grandmother, Susan Ned, boiled skunk cabbage and licorice fern roots and used the juice for a cold and cough medicine.[94] Interestingly, skunk cabbage root is noted by some contemporary herbalists as an expectorant, and it has an antispasmodic effect on the bronchial tubes.[95]

Other: Skunk cabbage makes a couple of appearances in Coos Bay mythology. The Little People built temporary shelters with skunk cabbage leaves and sticks.[96]

In Coos mythology, there is a cycle of stories about five generations of Tricksters who transformed the world in an ancient time before

Strawberry, *Fragaria chiloensis.*

human beings arrived and laid down customs that the Coos Bay people would follow. The first Trickster created salmon fishing technology and then offended the salmon. So he dug some skunk cabbage roots and learned to cook them properly by roasting them in hot ashes until they smelled sweet. Then he declared that the native people to come would do the same thing. He put the cooked roots in basket pans, and the Trickster created kinship terms. "Give this to your mother! your father!" and so forth until he had named all kinship terms.[97]

Technology: The leaves were used to collect a clay that was baked and mixed with fat to make a red paint.[98]

STRAWBERRY	Hanis: lǽləs
Fragaria chiloensis	Milluk: lǽləs

Plant description: Rosaceae, rose family. Perennial herb; can be 12 inches tall but often tends to grow close to the ground as a sprawling plant with numerous runners. Toothed leaflets are arranged in threes.

Flowers are white, with five to seven petals. Fruits are tiny strawberries. Grows in dunes and on beach margins and headlands.

Food: Wild strawberries are small, and they were not particularly sought out or preserved. When they were in season in midsummer, they were enjoyed as a fresh, sweet snack.[99]

SWEETGRASS Hanis: tł'ǘxwchu

Hierochloë occidentalis

Plant description: Poaceae, grass family. *Hierochloë occidentalis* is known as western vanilla grass or California sweetgrass. Stems are slender and up to 3 feet tall; leaves are smooth and flat, ¼ to just over ½ inch wide; flowers are in loose spikelets, olive to brown in color. Grows in forests from sea level to low altitudes west of the Cascades.

Food: Sweetgrass was layered above and below camas in earth ovens, as a spice. It was said to improve the flavor of the camas bulbs as they cooked overnight.[100]

Other: Indian women noted that pioneer women used this grass to make baskets, but Indians never did.[101]

TARWEED Hanis: yax

Madia sativa, M. gracilis, M. elegans

Plant description: Asteraceae, sunflower family. There are several species of tarweed in southwestern Oregon. They range in height from 2 to 6 feet. All have yellow disk flowers (sometimes the petals of *M. elegans* are dark red near the center of the flower) and mostly basal leaves, and the plants are sticky. They grow in disturbed areas.

Food: Several species of tarweed, of the genus *Madia*, grow in the Willamette, Umpqua, and Rogue Valleys but become uncommon west of the Coast Range, with the exception of *M. sativa*, coast tarweed. All have rich, oily seeds that were an important food for many Native Americans. In English, some native people began calling them "Indian oats," and they became known to many non-Indian settlers by the Chinook jargon name *sappolil.* Since tarweed is rare west of the Coast Range, it was an item of trade for coastal peoples.

The seeds are covered with a sticky, tarry coating. In late summer or autumn, the tarweed was burned to remove this coating on the seeds. The following day, women knocked the seeds into gathering baskets. Often the seeds were pounded into a meal, sometimes with other foods such as hazel nuts.[102]

Among the Upper Coquille, seeds of a sunflower, *Balsamorhiza deltoidea,* were gathered in a similar manner and often mixed with tarweed seeds. This was probably also done by other tribes who lived within the range of this sunflower.[103] The Upper Coquille also made a dish by soaking slices of pressed camas cakes in water, and then mixing this camas-sweetened water with mashed tarweed seeds into a kind of sweet soup or gravy.[104]

THISTLE

Cirsium edule

Plant description: Asteraceae, sunflower family. Edible thistle is a biennial or perennial herb with a stout taproot and hairy, reddish-brown stems up to 6 feet tall. Leaves are alternate and lance shaped, with spine-tipped lobes. Flowers are disk shaped and pinkish purple.

Food: Thistles, particularly *Cirsium edule,* were eaten in the Pacific Northwest. Some tribes ate the young peeled shoots. More commonly, the roots were eaten, usually cooked to break down the indigestible sugar inulin.

Lewis and Clark wrote about this root when they were camped along the lower Columbia River in the winter of 1805–1806. They noted that the crisp root resembled a parsnip and was about as thick as a man's thumb. When the thistle roots were roasted, they turned black and became sugary sweet, sweeter even than roasted camas bulbs.[105]

The Upper Coquille called thistle *łisalxwannœ'* and ate the peeled roots raw. Dorsey said the Alsea called it *pŭl-lí-cît-sló.*[106] Given its widespread regional use, it was almost certainly eaten by the Coos, Lower Umpqua, and Siuslaw as well.

THREESQUARE, AMERICAN
Schoenoplectus pungens

Hanis: c'ə́kwənł
Milluk: cəkwə́nł

Plant description: Cyperaceae, sedge family. American threesquare is also known as three-square bulrush, basket grass, and chairmaker's rush. It grows in brackish and freshwater marshes. The stems are triangular in cross section and grow up to around 4 feet tall. Each plant has two to four leaves, growing only on the lower third of the plant. Flowers are in spikelets near (but not at) the top of the stem.
Fiber: A plant described as a "three cornered marsh grass" that grew in salt marshes on the margins of the bay was gathered and made into skirts and baby clothes. The women's skirts of this "marsh grass" were regarded as strictly utilitarian and were worn for mucky jobs like clam digging. The plant was also used to make dresses for young children.[107]

TOBACCO
Nicotiana quadrivalvis

Hanis: dáha
Milluk: dáhai
Siuslaw: čıyúsan

Plant description: Solanaceae, nightshade family. The point of origin for *Nicotiana quadrivalvis* is probably southern California. Through trade, the seeds spread to the Pacific Northwest as far north as the Haida and Tlingit peoples, and east where it was grown by the Crow, Mandan, and Hidatsa.

Indian tobacco is a bushy annual herb that can grow up to 6 feet tall but is usually smaller. Leaves are alternate, with the longest leaves near the base of the plant. Flowers are white and trumpet-like. Plant is sticky to the touch.
Other: Tobacco was cultivated by native people in many parts of North America, including all of western Oregon. It was perhaps the only plant actively sown by seed in western Oregon, and most practices for growing it were similar in this region: A plot was burned to clear the ground. Tobacco seeds were scattered in the plot and brush fences were placed around it to protect it from wind. When the plants were mature, the leaves were harvested, dried, parched over flame, and stored in pouches.

Dried leaves of kinnikinnick were mixed in. The kinnikinnick was said to smell good and improve the taste of the tobacco and make it cooler.[108]

The stems and seeds were left in the plot to help reseed it. Sometimes seeds of wild-growing tobacco were collected as well.[109]

In southwestern Oregon, only men smoked when they gathered at meetings or to socialize in the evenings. The only exception was that some women doctors smoked as part of their curing ceremonies, although this was generally unknown among Coos, Lower Umpqua, and Siuslaw women doctors.

Tobacco was smoked only in the evening because it had powerful effects. Sometimes smokers would fall over and tremble or act stupefied. In Hanis, the term *ts'si* describes this state. Other men would laugh when this happened and say, he drew too hard. If two headmen were at a meeting, each headman passed a pipe around the assembly, and each man present smoked from the pipes as they were passed to him. If a man refused to smoke, it was a very bad sign. It indicated enmity, and this man would not be trusted by the others.

Most of the pipes were a straight tube, with a wooden stem that could be made from a variety of local woods. The pipe bowl itself could be made from drilled and polished sandstone or clay, or it could be carved from steatite.

There were other customs surrounding tobacco. Smoke was blown in the face of male babies. During thunderstorms, tobacco along with paint, paddles, and fishnets were put in a fire. It was said that the Thunders, chiefs of the ocean and its denizens, grew angry when their beloved salmon were mistreated. Any mistreatment of salmon could cause a thunderstorm. People said, "Thunder, go on north, they are abusing your children there."[110]

TULE
Schoenoplectus tabernaemontani, S. acutus

Hanis: c'lǽpəł, c'ıllǽpıl
Milluk: c'ə́ llæbənł
Siuslaw: čítłq

Plant description: Cyperaceae, sedge family. Soft-stem bulrush (*Schoenoplectus tabernaemontani*, formerly *Scirpus validus*) and hard-stem bulrush (*Schoenoplectus acutus*, formerly *Scirpus acutus*) are very similar in appearance, and both are locally called tules. Both are 3 to 9 feet tall, grasslike and round in cross section, with flowers in brown spikelets at the top of the plant. *S. tabernaemontani* seems to be more common in the upper parts of estuaries where the water is more fresh than brackish, and *S. acutus* is common in freshwater lakes.

Fiber: Mats made of tules (*číšıl* in Hanis and Milluk, *pılk* in Siuslaw) were useful in many aspects of daily life. The tules were gathered in summer (late in the season they became speckled with black spots and were generally no longer harvested then), dried, and then sewn together into mats with large vine maple or deer bone needles and cattail twine. Sometimes mats were decorated with extra twined elements. Mats were laid down on the floors of houses, hung on walls as extra insulation, and used as bed mats or seat cushions in canoes.[111]

A large soft storage basket, *bínı* in both Coos Bay languages, was made by folding a tule mat and sewing up the sides with cattail thread, iris leaf thread, or conifer roots. It was used to store food, from nuts to dried meat, and could be laced shut.[112]

After a woman gave birth, the placenta was placed in a bag that was constructed similarly to a *bínı*. It was then placed up in a tree.[113]

Another versatile use of the tule mat was to make a sleeping bag. These were useful for hunters or others who were traveling far from home and needed to set up camp. Two large mats were laced together, leaving the upper part open. A person could crawl inside with an elk hide for warmth and lace up the top, safe and snug inside the tule sleeping bag. This same sleeping bag was also used during the coldest winter nights in plank houses.[114]

Tule floats, called *pápa* in Hanis and Milluk, were used in fishing. Annie Miner Peterson gave a good description of the making and use of the *pápa*:

> Perch were caught with hook and line in the bay. Flounders too were caught with hook and line. Some hooks were wood, some bone, of 2 pieces, V shaped, baited. Halibut and salmon were caught in the ocean that way. For bay fishing with hook and line they made a float [Hanis, Milluk] *pá•pa* of tules tied tightly together; 2 ft lengths; then bound round with limber

hoops, perhaps hazel switches. The line was attached to this float, and it was about 10 ft or so long, with a hook on the end. You follow your float, if a fish is pulling it, in a canoe. They get mostly Chinooks, last part of June when the Chinooks come in, July. The [Gregory Point] lighthouse people did this fishing in the ocean took for Chinooks. Small fish like shiners, herring, sardines etc are used for bait. Each person can identify his own float. Mrs. P's mother fished this way, but Mrs. P suspects that in prewhite days the women did not go out for the floats, only the men, and that the women cut up the fish.[115]

WAPATO
Sagittaria latifolia

Hanis: kwí'məc, qwí'mɪc
Milluk: qwí'mɪc
Siuslaw: qwí'mɪc

Plant description: Alismataceae, water-plantain family. Wapato, also known as arrowhead, grows at the edges of rivers, ponds, lakes, and ditches. Usually the plants are partially submerged but they can also grow in exposed mud. Plants grow from underground tubers that resemble small potatoes. Leaves are up to 12 inches long and resemble large arrow points. Flowers are white and bloom in summer.
Food: The tubers of this plant were gathered in the shallow waters of lakes and riverbanks. Mostly women harvested the tubers, by tromping in the mud and letting them float to the surface. Sometimes men helped with the harvest. The tubers were roasted in ashes and eaten fresh or after having been dried for winter storage. Wapato was often accompanied by salmon eggs or seal oil.[116]

When potatoes were introduced, they were given the same name as wapato.[117]

The outlet of Tenmile Lake, Tenmile Creek, was the boundary creek between the Hanis and Lower Umpqua peoples, and Tenmile Lake was mentioned as the usual gathering place for these tubers. Probably both Lower Umpqua and Hanis people met around these lakes to collect wapato, and language contact may account for the similarity of the names for this plant in Hanis and Siuslaw.

Wapato was probably gathered in a few other wetlands as well—historically some was reported in Siltcoos Lake, and recently tribal members discovered a large patch in the Umpqua River, as well as a small population in an unnamed pond just north of Tahkenitch Lake. A few wapato from the unnamed pond were transplanted successfully to a pond on the edge of the Tribal Hall on the reservation in Coos Bay.

YERBA BUENA

Hanis: čílčílə łǽłəx

Satureja douglasii
(Clinopodium douglasii)

Plant description: Lamiaceae, mint family. This is a ground-hugging vine with a classic mint aroma. Leaves are opposite, oval to nearly round, with short stems. Flowers are white and trumpet shaped. This plant is also known as mountain tea and Oregon tea.

Food: The leaves and vines of this mint were gathered in summer, dried, and stored to make tea anytime.[118] It is still gathered for tea today.

Medicine: The tea was also believed to be medicinal. The name translates as "pain's medicine." The *číłčıl* was a type of poison pain power that native doctors could send to hurt, or even kill, a victim. Yerba buena tea could help treat this, and it was a pleasant tea as well.[119]

8 Ferns, Fern Allies, and Moss

FERN, BRACKEN
Pteridium aquilinum

Hanis: łkwa (rhizome), łk'wítɪmł (plant, ferns in general)
Milluk: łq'wa (rhizome), łq'watímł (plant)
Siuslaw: yáuxa (rhizome)

Plant description: Dennstaedtiaceae, bracken family. Tall fern, grows up to 4 feet tall, in partial shade to full sun. Fronds are divided in threes, pinnately compound.

Food: The large rhizomes of bracken fern were widely used as food in the Pacific Northwest. They were harvested in summer and fall, and perhaps in late spring as well. The rhizomes grow deep in the soil and can be hard to dig up, so groups of women often worked together and sang songs. Usually the women gathered only rhizomes that were at least 2 feet long. The rhizomes were packed in pack baskets and taken home to be stacked near the fire to dry for a few days. Then the rhizomes were cooked—either directly in the fire or in an earth oven. After being cooked, any charcoal was scraped off and the rhizome pounded with a maul. The "bark" was scraped off and discarded, as was a stringy core inside the rhizome. The pulp was eaten with salmon eggs.[1]

Lottie Evanoff observed that pioneers cooked the fiddleheads and often ate them with cream gravy, but Indians traditionally did not eat the fiddleheads, only the rhizomes.[2]

The leaves of bracken and sword ferns were also layered into earth ovens for cooking roots or meat.[3]

Other: The terms *ɬk'wítɪmɬ* and *ɬq'watímɬ* were also used as a general word for any type of fern. Fern leaves were used to help clean eels and salmon, and slabs of meat or fish were placed on piles of fern leaves.[4] Usually the leaves of sword fern were used for these purposes, but sometimes so were those of bracken fern and probably some of the other local ferns as well.

FERN, LICORICE

Polypodium glycyrrhiza

Plant description: Polypodiaceae, polypody family. Small fern, single frond with pinnately compound leaves, grows up to 1.5 feet tall in shaded habitats. Often found growing on logs, trees, and rocks.
Medicine: The Upper Coquille people chewed the rhizome of the licorice fern as a cough medicine. It was also enjoyed when drunk with water because it tasted good.[5]

Beverly Ward, writing about her husband's Lower Coquille grandmother, Susan Ned, said she boiled skunk cabbage and licorice fern roots and used the juice to treat colds and coughs.[6]

The use of licorice fern as a medicine for coughs or sore throats is so well known throughout the Pacific Northwest that the Coos, Lower Umpqua, and Siuslaw peoples almost certainly used it as well.[7]

FERN, MAIDENHAIR

Adiantum aleuticum

Hanis: č'æ'yǽnæ ɬk'wətímɬ ("fine fern")

Plant description: Pteridaceae, maidenhair fern family. The maidenhair fern is also known as five-finger fern. Grows up to 1.5 feet tall on a long black stem, topped with several fronds arranged palmately. Usually grows in fertile, moist soil near streams.
Fiber: The black "bark" of the maidenhair fern stem was used by many weavers of southwestern Oregon and northwestern California as a black overlay in basketry. The cross section of the stem is somewhat oval. One side of the stem is shiny black, the other half reddish. The reddish side is more brittle than the shiny black side. In traditional basketry, weavers

from the coast of southwestern Oregon and northwestern California usually worked only with the shiny black stems, although occasionally a Yurok or Karuk weaver would mud-dye the reddish stems. This process made them not only black, but also less brittle.[8]

Traditionally, the Coos, Lower Umpqua, and Siuslaw were aware of this use of maidenhair fern, but basket makers generally preferred to use eelgrass (*Phyllospadix scouleri, Zostera marina*) instead. However, many contemporary weavers work with maidenhair fern, preferring its shinier black over the duller black of eelgrass.

FERN, SWORD — Hanis: č'æš
Polystichum munitum — Milluk: č'æš

Plant description: Dryopteridaceae, woodfern family. Common understory plant of coniferous forests. Fern grows radially from a round base, resembling a clump of single fronds. Fronds can grow up to 6 feet high, although most are close to 3 feet or a little less.

Technology: Sword ferns were used to wipe up slime and blood when gutting fish. When a deer or elk was being butchered, sword ferns were laid down as a mat on which organs or strips of meat were placed to keep them clean.[9]

Strips of smoked meat or salmon were piled up in layers about 2 feet thick on top of a layer of ferns, with more fern leaves laid on top, and women would stand on it and mash the meat underneath to make it more tender.[10] Sword ferns were also put into earth ovens, around the roots or meat to be cooked, along with other ferns and herbs.[11]

Along with grasses and unspecified "weeds," sword ferns covered the women's sweat lodge.[12] The fronds were also tied to sticks and used as walls for temporary shelters. They could also be used as roof thatch for these same shelters, but bulrush (*Scirpus microcarpus*) was more often used for the roof.[13]

HORSETAIL
Equisetum sp.

Hanis: məkáwa, káwaa

Plant description: Equisetaceae, horsetail family. Horsetails are from an ancient lineage of lower vascular plants. When they first emerge from the ground in spring, they resemble asparagus stalks. Soon they leaf out with stiff, needlelike leaves and resemble a bottlebrush.
Food: In April, the upper 2 to 3 inches of the stalks are tender and edible, and these were broken off to be taken home and eaten raw.[14]
Other: There was a village, *Nıkkáwwáha*, a short distance down the bay from Empire that was named after this plant. This was the northernmost village of the Milluk people.[15]
Technology: Horsetails contain silicates and were used as sandpaper to put a finishing polish on wooden bowls and the shafts of fishing poles.[16]

MOSS

Hanis: gwaht
Milluk: hánabant'
Siuslaw: qut

Plant description: Mosses make up the phylum Bryophyta. They are nonvascular plants (meaning they lack certain specialized tissues that most land plants have to transport water and minerals) and so are usually quite small, 4 inches or less in height. They are quite common in damp western Oregon, growing on trees, logs, rocks, lawns, homes, and practically any other outdoor surface one can think of.
Medicine: Sphagnum mosses were wrapped on wounds to absorb blood and help healing.[17]
Other: The Hanis and Milluk word for moss was the same word used for tripe because it looked greenish before it was peeled.[18]

Leaves or moss were used to clean oneself after defecation.[19]

Technology: Mosses were sometimes used as a kind of rag to clean blood and slime while gutting fish. Smaller fish such as flounder, herring, and smelt were run through moss to help remove the scales in preparation for drying or cooking them.[20]

The species of moss that grow on bigleaf maple (*Acer macrophyllum*) were pulled off in large flat sheets. The inner surface of this moss mat was smooth. For tanning a deer hide, the brains were placed on the mat of moss and tied up in a bundle, until the tanner was ready to work the brains into the hide. The mosses that grow on alders were also used for this but did not work as well as the maple tree mosses.[21]

Dried moss was used to start fires with a fire drill. When traveling, people packed a coal wrapped in red cedar in a basket of moss, with the outer layer of moss dampened so it would not burn through.[22]

9 Fungi and Seaweeds

FUNGI, SHELF Hanis: gwǽsk'wɪs
Basidiomycota

Plant description: Shelf fungi belong to several families in the phylum Basidiomycota. They all tend to be firm to the touch (although some are soft) and grow in a vague fan shape without stems directly out of logs and snags.

Other: A shelf fungus, described as being white on one side and brown on the other, was said to cause echoes. People who mocked and teased were also called *gwǽsk'wɪs*.[1] This is the Milluk word for "echo" and may also be the word for shelf fungus in that language.[2] This belief is not unique to the Coos people. The Quinalt also believed that shelf fungus caused echoes because of its ear-like shape.[3]

The turkey fan tail, *Trametes versicolor*, is a common shelf fungus in North America. *Gwǽ'skwɪs* probably included this species and many related species that grow in western Oregon.

Interestingly, even though numerous species of edible fungi are indigenous to the region, none were mentioned as part of the traditional diet. Lottie Evanoff could recall no cultural uses for mushrooms, other than playing with them when she was a child.[4]

But postcontact, some Indians did begin to pick, and often sell, mushrooms. Frank Drew said, "In September the Florence sand dunes where pines grow are just full of mushrooms and [they pick] them. Just as soon as the ground gets soaked, mushrooms start coming up."[5]

KELP
Nereocystis luetkeana, Alaria marginata, Macrocystis sp.

Hanis: qálaqas, qalqas
Milluk: qaləqas
Siuslaw: páhu

Plant description: Phaeophyceae, brown algae. Several species of kelp regularly wash up on beaches. They are easy to recognize by their long stipe (the equivalent of a stem), and bladder topped with leafy blades. Kelp forests are important areas of biodiversity, sheltering many species of fish and other marine organisms.
Other: Children sometimes played with kelp that washed up on beaches. Kelp beds were thought of as sea otter beds.

The Chinese community of Coos Bay boiled kelp with sugar to make candy.[6]

Some Pacific Northwest tribes used the long stalks of kelp to make fishing line.[7]

SEA LETTUCE
Porphyra perforata

Hanis: tɬ'kínɪx
Milluk: tɬ'kínɪx

Plant description: Rhodophyta, red algae. This algae grows as a "sheet," with color ranging from light red to purple and a smooth, shiny surface. Red sea lettuce grows on rocks from fairly high up in the intertidal to the subtidal zone, ranging from Alaska to Baja California.
Food: This seaweed was gathered from the rocks, taken home, and laid out on mats or boards to dry in the sun. The dried seaweed was kept in storage baskets, ready to be eaten anytime. It was usually dipped in seal oil, whale blubber, or elk tallow for "flavoring," according to one informant.

Darker-colored plants were preferred over those of a lighter red. The *tɬ'kínɪx* was probably *Porphyra perforata*, and perhaps the closely related species *P. lanceolata* as well.[8]

Frank Drew thought that the Coos Bay people had adopted eating seaweed from the Rogue River people. However, this seems unlikely—seaweed was also eaten by the Alsea and Tillamook peoples.[9]

The local Chinese community, in the late nineteenth and early twentieth centuries, used to gather a lot of this seaweed as well.[10]

10 Unidentified Plants

There are several plants mentioned in various ethnological field notes that are not described in enough detail to identify.

FOOD PLANTS

Shore lupine (*Lupinus littoralis*) is a prized root food gathered in the sand dunes, but there were two other sand dune plants, as yet unidentified, that were also gathered for their edible roots.

A plant known as *tłəmqáá'yawa* in Hanis and Milluk was described by Frank Drew as "a sort of turnip, that grows in pure white desert sand til they look (the leaves) like a watermelon plant. The fruit is similar of beanpods [*sic*], but smaller. The roots is [*sic*] dug out. . . . The roots are not cooked, but are eaten raw. The root exterior is torn off, chewed and chewed, and that is all that is done to this wild turnip. It tastes like candy, sweet."[1] Annie Peterson mentioned this same plant only briefly, noting that it was a "turnip" that grew in marshes.[2] There are only a few plants that grow in the Oregon dunes that have edible roots, so the possibilities are limited. The description of watermelon-like leaves and small seeds or seedpods resembling bean pods does not clearly match any of them. However, indigenous descriptions of leaf shape are often not helpful. One informant said that springbank clover leaves resembled those of a carrot, although there is no particular resemblance other than that the leaves of both plants could be described as divided.

The most likely candidate is the yellow sand verbena, *Abronia latifolia*. It grows in the dunes (before widespread growth of European beach grass, it was formerly common in the dunes), produces seedpods that are up to half an inch long, and has large sweet-tasting roots that are edible raw.[3]

Yellow sand verbana showing part of its large root, *Abronia latifolia.*

Another hint is in the etymology of the name. In Hanis, the suffix -*ayawa* is a nominalizing suffix usually added to verbal stems. The verb root *tl'imq* or *tlimaq* means "to have a scent, to stink." Thus *tłəmqáá'yawa* means "stinker, scented." One of yellow sand verbena's distinctive features is the strong sweet smell of its flowers.[4] Very few native Oregon dune plants have flowers with a noticeable scent.

A more remote possibility is a plant that Scottish botanist David Douglas collected along the lower Columbia in the summer of 1825 that he classified as part of the pea family in the genus *Lathyrus.* He said it was a perennial plant with "flowers large bluish-purple, a splendid strong-growing plant; the roots are large, run deep in the sand, and are eaten by the natives in a raw state; abundant."[5] Yet no subsequent ethnobotanist or wild food crafter has reported the roots of any native North American species of *Lathyrus* pea to be edible. The identity of Douglas's *Lathyrus* is uncertain. He may have confused the roots of one plant with those of another, which was the opinion of three nineteenth-century writers.

George Suckley and James Cooper, on a botanical and zoological survey for the railroads in the 1850s, thought David Douglas had mistaken the roots of *Lupinus littoralis* (seashore lupine) for those of yellow sand verbena. They noted that "*L. littoralis*, Dougl., somewhat resembles this, but I met with none of which the roots were used by the Chenooks [*sic*] as food. They do dig in the same place the roots of an *Abronia* he may have mistaken for those of lupine."[6] Scottish botanist Robert Brown, writing of his trip to North America in the 1860s, noted the same: "Douglas says that the roots of *Lupinus littoralis*, Dougl., are eaten by the Indians near the mouth of the Columbia River (Chinooks). I never knew them to do so but I have seen the natives at the same place eat the roots of *Abronia arenaria*, Menz. [an old synonym for yellow sand verbena], which he might have mistaken for the former plant."[7] If Suckley, Cooper, and Brown are correct, then the Coos dune "turnip" *tłəmqáá'yawa* was very probably yellow sand verbena.

The third dune plant was mentioned briefly by only one Coos Bay informant, Agnes Johnson. Called *takus* in Hanis, it was said to be found in the dunes, and its root was eaten after roasting in earth ovens.[8] A likely possibility is Pacific cinquefoil, *Potentilla anserina*. Pacific cinquefoil was a root food widely utilized by Pacific Northwest coast tribes, and it was almost always roasted because the raw roots are bitter. It is common in the sand dunes, but without any further description, it is impossible to be certain.

There was a marsh plant, known as *kw'yais* in Hanis, with roots that were edible after being baked. A second marsh plant, called *qlutł* in Hanis, was specified as a grass (or with grasslike leaves) and had roots that were edible raw.[9] Several plants that can be found in marshy habitats have edible roots, including some species in the rush family (Cyperaceae) and some liliaceous plants with edible bulbs, such as the tiger lily (*Lilium columbianum*) and onions (*Allium* sp.).

Jim Buchanan said there was an edible wild "parsnip" called *šámaš* in Hanis.[10]

Lottie Evanoff said that when she was a child there was an edible wild parsley that grew along the bluffs in Yachats.[11] Parsleys are part of the family Apiaceae (formerly Umbelliferae), and there are several species with edible roots. Given the habitat, this species is probably *Conioselinum gmelinii*, Pacific hemlock-parsley, which grows on beaches and coastal bluffs. This plant is the "Indian carrot" of the British Columbia coast. The patches were marked in summer and the roots dug in spring before the leaves came up. The roots were steamed and eaten.[12]

MEDICINAL PLANTS

Frank Drew claimed there was a tree or shrub that grew inland whose bark made a tonic tea. The tree and medicinal tea were both known in Hanis as *ník'ɪn šáhɪ*, the wood drink. Women traveled to the woods, gathered the bark, dried it in the house rafters for a long time, and then stored it in sacks until ready to use. The seasoned bark was soaked in cool water until the water turned yellow. The bark was discarded, and this tonic was given to babies occasionally to prevent stomach troubles. Adults drank a little tonic every few days in the belief that it would keep them healthy. The tonic was described as bitter. In the Siuslaw River area, this tree or shrub grew as far west as Mapleton.[13]

A plant with wide leaves was called "swelling medicine," *pxwánəsœ łœ́łəx* in Hanis and *pxwánəs dœ łœł* in Milluk. The leaves of this plant were heated and then placed on swollen joints or muscles. Melville Jacobs thought it might be a plantain and took a sample, although I do not know whether this sample was ever identified.

"Wild lettuce" was used in a similar way to "swelling medicine" and was also drunk as a tea. This could have been miner's lettuce, *Claytonia perfoliata*, which was used as a poultice for cuts, or a closely related species, *C. sibirica*.[14]

"Blistering medicine," *c'ɣət'œ́his łœ́łəx* in Hanis, was described by Annie Miner Peterson as a "tiny plant [that grows] in moss, has 2 leaves sticking up. It was used as a sort of plaster. The women get it, dry it, and is used above a pain. It is very strong and blisters and it draws out the pain and disease. . . . It makes a yellow matter come out. It is somewhat poisonous if put in mouth though it doesn't kill."[15]

The bark of a shrub, about 2.5 feet tall, that grew in rocky places up the Coos River was given to children to chew for sore throats. Lottie Evanoff described the bark as being about as thick as cinnamon bark.[16]

TECHNOLOGY AND FIBER

A plant called *gwałgwɪ* in Hanis and Milluk was used to make a strong thread for sewing garments. The plant was said to be short and "flowerless"; the leaves were pulled in summer, dried, and then worked and twisted into thread.[17]

The "tiny parsnip or potato like roots" of an unidentified plant were cooked in ashes and then rubbed on water baskets to make them more watertight.[18]

There was a light green plant resembling a small tule, eaten by cattle and called *dúggwa* in Hanis, that grew thickly enough at Yarrow Beach (now part of North Bend) that the village there, *Dúgwahaich* or *Duwœtɪch,* was named for this plant. It also grew across from Empire. The plant was heated with fire and used for tying bundles.[19] This may be a species of *Eleocharis,* a variety of sedge.

A plant described as a "white bladed marsh grass," known in Hanis as č'ínəm, was wrapped around the end of shinny clubs as part of the steaming and shaping process.[20]

OTHER

Lottie Evanoff remembered running and playing as a child with Alsea children at Yachats. The Alsea children ran through patches of a "sticker grass" that she called *hwás-hwi.* Lottie could not run through it; she found it too uncomfortable. This sticker grass was about 8 inches tall.[21]

Appendix: Basketry

Now a seal came from out of the water.

"Hey, go away!" (said the Trickster)
It wouldn't go away.
It came up close.

When it came up again, he threw the *paštala* hat at [the seal].
It dropped right on its head.

"When the next people are here and see you, you will never hurt anybody.
Your head will be just like the *paštala*."

Now then a whale emerged and spouted.
"Get away! I'm afraid of you."

He took the basket,
it was rough edged and unfinished, that basket.

He threw [the basket] in its mouth.
"You will never bite anybody.
You will have no teeth,
but they will all be stuck up inside.
You will never do anything to anybody."

—"The Trickster Person Who Made the Country"[1]

Coos Bay basket 2-13213, Phoebe Hearst Museum, Berkeley, CA.
Annie Miner Peterson identified it as possibly a *gıdamən*-type basket that could be used for cooking.

All Coos, Lower Umpqua, and Siuslaw baskets were made by a method of twining, where two or occasionally three elements (the woof, or weft) were woven with a half twist around spokes (the warp). Some baskets were twined wholly of softer materials, such as cattail leaves or tules, while other baskets used more rigid materials, such as the roots of conifers or hazel switches, for either warp or weft, or both. Less frequently, willow and spruce sticks were used for the warp. Geometric designs were woven in using other plant materials that were sometimes dyed red, yellow, brown, or black—beargrass for white, eelgrass for black, cherry bark, maple bark, cedar bark, cattail leaves, and many other materials.

Women were the primary weavers, although men did work on fish trap baskets and nets. Girls accompanied their mothers and other women to help gather and process basket materials, and they learned to weave by watching and helping. They learned what plants were useful, where to find them, what time of year to collect them, and how to prepare the materials for weaving.

Baskets were so important and useful in many facets of life that in the Hanis and Milluk languages there were more than two dozen terms for them. They are listed in the following table. Most of the terms were given by Annie Miner Peterson, a Coos weaver. Siuslaw terms came from Louisa Smith, Spencer Scott, and Frank Drew.

HANIS	MILLUK	SIUSLAW	DESCRIPTION
Storage baskets			
táutau			Smaller to medium-sized basket that can be used for storage, packing, and berrying.
mígæ	mígæ		Large food storage basket, open top woven with hazel and spruce roots. It is larger than a packbasket.
gwə́nɪ	gwə́nhɪ		Large undecorated soft storage basket made of tules to store roots and nuts in the house.
binɪ			Large soft tule basket used for food storage. A tule mat was folded in half and the sides were sewn up with conifer roots, *Juncus* rush, or cattails. Basket is open at the top but could be laced shut.
axu			Basket for storing valuables.
dáasəc’	dáasəc’		Soft basket for storing valuables, made of cattails and decorated with overlays of eelgrass, beargrass, and alder-dyed cedar bark, with a handle or pack rope. At one point Mrs. Peterson described it as a kind of handbag.
qǽtłætł			Finely made basket, can be used to store valuables. Similar to *axu*.
čɪllał	čɪláał		A round basket about 1 foot high, 1 to 2 feet wide. Soft, made of cattail, worked with designs of beargrass, cedar bark, and eelgrass. Laced shut at the top, has a handle.

HANIS	MILLUK	SIUSLAW	DESCRIPTION
bʊ́'ʊs	búw əs		Purse for dentalium, made of glossy hazel sticks, up to 8 inches long, 4 wide. The same word was also used for purses carved from elk horn.
		q'áuq'áuni	Basket for storing valuables.
xʊkwæ, hʊkwǽ	xʊkwǽɬ, hʊkwǽl		Water storage bucket made of red cedar roots; holds about 1 gallon of water. Braided handles also of red cedar roots.
pílč'æ	pílč'æ		Larger water bucket, also made of red cedar roots, with a wider mouth at the top than *hʊkwǽ/ hʊkwǽl*. Could be carried by two people with a stick through the handle.
qa'lánč	qa'lánč	qalanč	Larger and taller than the *qǽlæ'æn*, used to store dried berries. Made of conifer roots and cattail leaves.
Pack baskets			
káwəl	k'ha	káwəl	Conical pack basket, almost *V*-shaped. This was the most commonly used pack basket. Warp made of sticks, weft of spruce roots, with pack ropes woven from rushes.
k'áwəl	ts'ǽwæl		Conical pack basket similar to the *káwəl / k'ha*, but with a smaller-diameter top opening.
cə'ɣǽiɬæ	cə'ɣǽiɬæ		Large, stiff, *V*-shaped carrying basket made of spruce limbs or hazel switches woven with spruce roots. Larger than a *káwəl* burden basket, used for heavy work like gathering firewood.

HANIS	MILLUK	SIUSLAW	DESCRIPTION
pq'á ɬ	pq'áá ɬ		Lower Umpqua–style pack basket, *U*-shaped rather than conical, woven with cattail leaves.
Trap baskets			
hák	hák	cú'un	Conical fish trap basket up to 14 feet long. Often made of young split fir sticks woven with spruce or fir roots, with a vine maple hoop.
glúbət		wap	Fish trap basket. Conical, smaller than the *hák* fish basket. Made of hazel switches and vine maple hoops. Used in upriver fishing sites for lamprey.
tæc'	tæc'		Trap basket for Dungeness crab; flat bottom, round sides.
		skwíltɬa	Salmon basket. Size and appearance not described.
pápa			Floating buoys for fishing lines and traps, made of tules.
Clothing			
tɬp'ə́ la	tɬp'ə́lla, šap'ála	ɬkwánukw, ɬkwálukw	Coos Bay–type women's basket hat. It is a little taller than the northwestern California–style cap and is usually made of cattail leaves, sometimes with some *Juncus* rush added in, decorated with beargrass for white and eelgrass for black designs.
paštála	paštála		Northwestern California and southwestern Oregon–style women's cap, usually made with willow or hazel sticks for a warp, and conifer or willow roots for a weft.

HANIS	MILLUK	SIUSLAW	DESCRIPTION
Other household items			
tł́æxæč	tł́æxæč		Openwork basket made of roots or rushes
qæ'lǽ'æn	qæ'lǽ'æn		Berry-picking basket made of fir or spruce roots. A rope was attached to the basket and hung from the neck (rope made of cattail leaf or *Juncus* rush).
mádan, maadán	maadán		Grinding basket with an open bottom. It was placed over a mortar, and food such as acorns or manzanita berries was pounded with a pestle.
bǽsɪk	pásɪk'	k'anɪšk'	Small dipper or water cup. Made of red cedar roots, sometimes with two hazel switches tied to one side for a handle.
gidámən			Could be used as a cooking basket. Made of conifer roots.
č'šɪl	č'šɪl	pɪlk	Tule mat, usually sewn together with cattail-leaf twine.

Notes

PREFACE

1 Frachtenberg 1913:8–9.

CHAPTER 1

1 Frachtenberg 1909; M. Jacobs 1932–1934, 91:34; US Court of Claims 1931.
2 Frachtenberg 1913, 1914; M. Jacobs 1939; Grant 1994a, 1994b; Hymes 1966; Mithun 1999.
3 Aikens, Connolly, and Jenkins 2011; DeLancy and Golla 1997; Grant 1994a, 1994b; M. Jacobs 1939; Mithun 1999.
4 Aikens, Connolly, and Jenkins 2011; E. Jacobs 2003; Mithun 1999.
5 Jacobs 1932–1934, 93:7, 20; D. Whereat 2011:32.

CHAPTER 2

1 Hymes 2003:327.
2 Frachtenberg 1909; M. Jacobs 1939:84–90, 118–20.
3 Frachtenberg 1909; M. Jacobs 1932–1934, 91:43–44, 51, 101:24.
4 Harrington 1942, 24:92.
5 M. Jacobs 1932–1934, 91:7–9, 93:19; Harrington 1942, 21:908, 22:1022a, 1176, 24:214a.
6 Hines 1852:103–4.
7 M. Jacobs 1932–1934, 91:33, 132; Harrington 1942, 22:740a, 24:546a.
8 Harrington 1942, 22:814b.
9 Harrington 1942, 24:243b; M. Jacobs 1932–1934, 91:17.
10 Harrington 1942, 24:153b, 194a, 196b, 759b.
11 Beckham et al. 1984; D. Whereat 2011:261–66, 326–29.
12 D. Whereat 2011:326–29.
13 DuBois 2007.
14 Beckham et al. 1984; D. Whereat 2011:326–29.
15 Jacobs 1932–1934, 95:43–47.
16 M. Jacobs 1932–1934, 92:149.
17 Beck 2009; Beckham 1986; Dillon 1975; Dodge 1898.
18 Harrington 1942, 24:936a.
19 Beck 2009; Beckham 1983; Bensell 1959:148; Tveskov 2000:467–70; US Court of Claims 1931.
20 Drew 1858.
21 Geary 1860.
22 Frachtenberg 1909.
23 M. Jacobs 1939:106.
24 Schwartz 1991:231.
25 Harrington 1942, 22:811b.
26 Harrington 1942, 24:257b.
27 Odeneal 1873.
28 *Annual Report of the Commissioner of Indian Affairs* 1862:299.
29 Beckham 1983.
30 Beck 2009; Beckham 1983.
31 Harrington 1942, 24:975b.
32 Beck 2009:100; Beckham 1983.
33 Beck 2009:106; Beckham 1977:182.
34 Beckham 1977:182.
35 Beck 2009; Beckham 1977.
36 Beck 2009:158–60; Beckham 1983.
37 Beck 2009:166–67.
38 Beck 2009:207–8.

CHAPTER 3

1 Coyote to the first people, Frachtenberg 1914:33.
2 Dorsey 1884; Grant 1994a, 1994b.
3 Beckham 1969.
4 Harrington 1942, 24:273b, 24:249ab, 348a; M. Jacobs 1932–1934, 91:79.
5 Frachtenberg 1909; Harrington 1942, 24:887a.
6 Beckham 1983; Frachtenberg 1909; M. Jacobs 1932–1934; St. Clair 1903; US Court of Claims 1931.
7 Frachtenberg 1909; St. Clair 1903; D. Whereat 2011.
8 Harrington 1942, 23:386a.
9 Dorsey 1884; Frachtenberg 1914; D. Whereat 2011.
10 Grant 1994b; M. Jacobs 1932–1934; D. Whereat 2011:216–30.
11 M. Jacobs 1932–1934, 93:3, 5; M. Jacobs 1939:104–14; D. Whereat 2011; Youst 1997.
12 Drucker 1934; D. Whereat 2011:213–30.
13 Grant 1994a, 1994b; Harrington 1942; D. Whereat 2011:213–30.
14 Grant 1994a; D. Whereat 2011:213–30.

CHAPTER 4

1 Harrington 1942, 22:31b.
2 Schultz 1990; Wiedemann et al. 1999.
3 Anderson 2005:129.
4 P. Whereat 2001.
5 Barnett 1934.
6 M. Jacobs 1932–1934, 91:124.
7 M. Jacobs 1932–1934, 97:26.
8 US Court of Claims 1931.
9 Elizabeth Morrissey, personal communication, August 10, 2000.
10 Harrington 1942, 26:141.
11 Anderson 2005:194; Lalande and Pullen 1999:262–63.
12 Lalande and Pullen 1999.
13 Harrington 1942, 23:236a.
14 P. Whereat 2001.
15 Harrington 1942, 22:31b; Frachtenberg 1914.
16 Beckham 2006:265, emphasis added.
17 Moss 2005:276.
18 M. Jacobs 1932–1934, 98:96.
19 Maloney and Maloney 1933:34.
20 Harrington 1942, 24:572–73b.
21 Harrington 1942, 24:240, 1034a.
22 Moss and Erlandson 1995:61.
23 Harrington 1942, 24:303a.

CHAPTER 5

1 Frachtenberg 1914:82–83.
2 M. Jacobs 1932–1934, 93:40, 100:58–60.
3 Frachtenberg 1909; M. Jacobs 1932–1934, 91:62–63.
4 E. Jacobs 1935, 119:20.
5 M. Jacobs 1932–1934, 92:153.
6 Harrington 1942, 21:760, 905; Grant 1994b; M. Jacobs 1932–1934, 92:93.
7 M. Jacobs 1932–1934, 91:7.
8 Harrington 1942, 20:80a, 26:204.
9 Drucker 1934; M. Jacobs 1932–1934, 100:56–58.
10 Harrington 1942, 22:761, 26:260.
11 Gunther 1973:40.
12 Harrington 1942, 21:758, 22:772.
13 M. Jacobs 1932–1934, 91:63–65.
14 Harrington 1942, 22:1176, 25:796.
15 Moerman 1998:151.
16 M. Jacobs 1932–1934, 93:97–98.
17 Maloney and Maloney 1933:6.
18 Harrington 1942, 25:736.
19 Hines 1852:107.
20 M. Jacobs 1932–1934, 93:89–91.
21 Barnett 1934.
22 M. Jacobs 1932–1934, 93:89.
23 M. Jacobs 1932–1934, 91:132.
24 M. Jacobs 1932–1934, 98:28.
25 Frachtenberg 1909; Harrington 1942, 22:981.
26 M. Jacobs 1932–1934, 92:153; Harrington 1942, 21:910.
27 M. Jacobs 1932–1934, 98:82.
28 Frachtenberg 1909; Harrington 1942, 21:910.
29 Ward 1986:12–13.
30 Blackburn 2005:47.
31 Harrington 1942, 21:906a; M. Jacobs 1932–1934, 91:7–8.
32 Frachtenberg 1909; Harrington 1942, 24:446.
33 E. Jacobs 2003:238.

34 M. Jacobs 1932–1934, 93:30.
35 Harrington 1942, 22:1019a; M. Jacobs 1932–1934, 91:7–8.
36 M. Jacobs 1932–1934, 95:74.
37 Harrington 1942, 22:1050.
38 Harrington 1942, 22:781a.
39 M. Jacobs 1932–1934, 93:43.
40 M. Jacobs 1932–1934, 101:36.
41 Harrington 1942, 22:761, 26:260.
42 Harrington 1942, 22:801b.
43 Moerman 1998:428.
44 M. Jacobs 1932–1934, 101:23; Harrington 1942, 24:503b.
45 M. Jacobs 1932–1934, 91:36, 101:23; Losey et al. 2003.
46 M. Jacobs 1932–1934, 21:758.
47 Maloney and Maloney 1933.
48 M. Jacobs 1932–1934, 91:4.
49 M. Jacobs 1932–1934, 91:14, 92:109.
50 M. Jacobs 1932–1934, 100:14.
51 M. Jacobs 1932–1934, 96:82, 97:26.
52 M. Jacobs 1932–1934, 91:10, 92:85; Harrington 1942, 22:277b.
53 Maloney and Maloney 1933:9.
54 M. Jacobs 1932–1934, 100:14.
55 M. Jacobs 1932–1934, 92:31.
56 Ibid.
57 M. Jacobs 1932–1934, 92:29.
58 M. Jacobs 1932–1934, 92:160–62.
59 M. Jacobs 1932–1934, 100:46.
60 Maloney and Maloney, in M. Jacobs 1932–1934, 91:94.
61 M. Jacobs 1932–1934, 91:162.
62 M. Jacobs 1932–1934, 92:33.
63 M. Jacobs 1932–1934, 91:58; Harrington 1942, 23:488a.
64 Harrington 1942, 22:43, 24:276; M. Jacobs 1932–1934, 91:21, 92:89, 97:44.
65 Frachtenberg 1909.
66 M. Jacobs 1932–1934, 91:16.
67 M. Jacobs 1932–1934, 92:116.
68 Harrington 1942, 24:830a.
69 Harrington 1942, 24:736b.
70 M. Jacobs 1932–1934, 92:83.
71 Harrington 1942, 21:905b.
72 M. Jacobs 1932–1934, 92:133.
73 M. Jacobs 1932–1934, 91:125.
74 M. Jacobs 1932–1934, 92:99–100; Harrington 1942, 22:568b.
75 M. Jacobs 1932–1934, 91:14.
76 M. Jacobs 1932–1934, 92:113.
77 Harrington 1942, 21:750a.
78 M. Jacobs 1932–1934, 100:20–26.
79 M. Jacobs 1932–1934, 92:149.
80 Harrington 1942, 22:1036.
81 Harrington 1942, 22:487.
82 M. Jacobs 1932–1934, 93:40–41, 94:116.
83 M. Jacobs 1932–1934, 93:40–41, 100:58.
84 M. Jacobs 1932–1934, 93:101–2.
85 Barnett 1934.
86 Harrington 1942, 21:905.
87 M. Jacobs 1932–1934, 91:36.
88 Drucker 1934; M. Jacobs 1932–1934, 91:42; Harrington 1942, 25:191.
89 Douglas 1914:67.
90 Douglas 1914:225.
91 R. Hall 1984:42.
92 M. Jacobs 1932–1934, 100:20.
93 Drucker 1934; M. Jacobs 1932–1934, 100:124.
94 Harrington 1942, 26:153; Tveskov 2000:109.
95 Harrington 1942, 26:230.
96 M. Jacobs 1932–1934, 100:8.
97 Harrington 1942, 22:1226b.
98 M. Jacobs 1932–1934, 95:109, 100:142; Drucker 1934.
99 M. Jacobs 1932–1934, 95:109.
100 Harrington 1942, 24:164a.
101 M. Jacobs 1932–1934, 92:81.
102 Harrington 1942, 20:87.
103 Maloney and Maloney 1933:8.
104 Harrington 1942, 22:521; M. Jacobs 1932–1934, 93:94.
105 M. Jacobs 1932–1934, 93:100.
106 M. Jacobs 1932–1934, 93:43, 83.
107 M. Jacobs 1932–1934, 92:91–92.
108 M. Jacobs 1932–1934, 91:59.
109 M. Jacobs 1932–1934, 93:48.
110 M. Jacobs 1932–1934, 91:6.
111 Harrington 1942, 21:906a.
112 Frachtenberg 1909.
113 Barnett 1934.
114 Harrington 1942, 22:526a.

CHAPTER 6

1 M. Jacobs 1932–1934, 91:36, 101:23.
2 Dennis Rankin, personal communication, August 2, 2015.

3 Harrington 1942, 22:739b.
4 Drucker 1934; M. Jacobs 1932–1934, 91:36, 92:129.
5 M. Jacobs 1932–1934, 92:31; St. Clair 1903.
6 Harrington 1942, 20:63a; Kozloff 2005:320.
7 M. Jacobs 1932–1934, 91:37, 101:23.
8 Drucker 1934; M. Jacobs 1932–1934, 91:36; Harrington 1942, 24:178b.
9 M. Jacobs 1932–1934, 100:112.
10 M. Jacobs 1940:196.
11 Harrington 1942, 22:775b.
12 M. Jacobs 1932–1934, 91:37; Drucker 1934; Harrington 1942, 22:1200b.
13 Jacobs 1932-1934, 91:36, 92:118; Drucker 1934.
14 Harrington 1942, 22:764a.
15 M. Jacobs 1932–1934, 91:56–57.
16 M. Jacobs 1932–1934, 93:65, 97:118.
17 M. Jacobs 1932–1934, 91:46, 119–20, 92:122.
18 M. Jacobs 1932–1934, 91:11–13.
19 Ibid.
20 Harrington 1942, 23:287a, 288a, 290b; M. Jacobs 1932–1934, 91:11–13, 92:113.
21 Drucker 1934; Harrington 1942, 25:834.
22 M. Jacobs 1932–1934, 101:23.
23 Harrington 1942, 22:775.
24 Harrington 1942, 22:948; M. Jacobs 1932–1934, 93:94.
25 Harrington 1942, 24:751a.
26 M. Jacobs 1932–1934, 92:96.
27 M. Jacobs 1932–1934, 91:16; Drucker 1934; Harrington 1942, 22:87b.
28 M. Jacobs 1932–1934, 93:100.
29 M. Jacobs 1932–1934, 91:32.
30 M. Jacobs 1932–1934, 91:41; Harrington 1942, 26:177–80; Drucker 1934.
31 M. Jacobs 1932–1934, 91:56–57.
32 Harrington 1942, 26:141.
33 Harrington 1942, 21:908–9, 22:779.
34 M. Jacobs 1932–1934, 91:36; Harrington 1942, 22:54, 23:738b.
35 M. Jacobs 1932–1934, 91:64–65; Harrington 1942, 22:487.
36 M. Jacobs 1932–1934, 100:176.
37 E. Jacobs 2003:138, 190.
38 Drucker 1934; Frachtenberg 1920:224–26.
39 Harrington 1942, 24:731a.
40 M. Jacobs 1932–1934, 101:17; Harrington 1942, 22:794.
41 Harrington 1942, 22:794.
42 Harrington 1942, 22:584, 794.
43 Ward 1986:87.
44 Frachtenberg 1920:84.
45 M. Jacobs 1932–1934, 97:50.
46 Gunther 1973:43; Turner 1997:114–15.
47 Grace Brainard, personal communication, November 12, 2000.
48 R. Hall 1984:42.
49 Barnett 1934; M. Jacobs 1932–1934, 92:78, 93:96, 101:23; Drucker 1934.
50 Harrington 1942, 22:754.
51 M. Jacobs 1932–1934, 91:57.
52 Frachtenberg 1909.
53 M. Jacobs 1932–1934, 92:149.
54 M. Jacobs 1932–1934, 92:105.
55 M. Jacobs 1932–1934, 91:17, 37, 94:114; Frachtenberg 1909.
56 M. Jacobs 1932–1934, 92:74, 93:40.
57 M. Jacobs 1932–1934, 91:12, 92:113.
58 Frachtenberg 1909; Harrington 1942, 22:981b, 982a.
59 M. Jacobs 1932–1934, 91:25.
60 M. Jacobs 1932–1934, 92:135; Harrington 1942, 22:526a.
61 Frachtenberg 1909; M. Jacobs 1932–1934, 101:34.
62 Nelson 2000.
63 Harrington 1942, 24:791b; Drucker 1934.
64 Gunther 1973:30; Moerman 1998:329–30.
65 Harrington 1942, 24:791b.
66 Harrington 1942, 24:733b.
67 Harrington 1942, 22:805b; M. Jacobs 1932–1934, 91:66.
68 M. Jacobs 1932–1934, 100:32.
69 Moerman 1998:484.
70 M. Jacobs 1932–1934, 91:35; Drucker 1934.

71 Harrington 1942, 22:807a; M. Jacobs 1932–1934, 100:112.
72 Harrington 1942, 22:934a.
73 M. Jacobs 1932–1934, 91:33.
74 Ray Willard, personal communication, April 27, 2001.
75 Drucker 1934; M. Jacobs 1932–1934, 91:40, 101:14.
76 M. Jacobs 1932–1934, 91:40.
77 US Court of Claims 1931:191; Harrington 1942, 22:874a.
78 M. Jacobs 1932–1934, 101:5–7.
79 M. Jacobs 1932–1934, 99:178.
80 Frachtenberg 1909.
81 M. Jacobs 1932–1934, 91:40; Drucker 1934.
82 M. Jacobs 1932–1934, 92:67, 101:26.
83 M. Jacobs 1932–1934, 101:2–3.
84 M. Jacobs 1932–1934, 91:145.
85 M. Jacobs 1932–1934, 92:18.
86 M. Jacobs 1932–1934, 91:10, 92, 93:59, 101:26.
87 M. Jacobs 1932–1934, 91:11.
88 M. Jacobs 1932–1934, 91:49.
89 Harrington 1942, 22:487, 758, 23:903a; Maloney and Maloney 1933.
90 M. Jacobs 1932–1934, 91:36.
91 M. Jacobs 1932–1934, 92:159.
92 Harrington 1942, 22:791.
93 M. Jacobs 1932–1934, 91:24, 92:92.
94 M. Jacobs 1932–1934, 91:24.
95 M. Jacobs 1932–1934, 92:100.
96 M. Jacobs 1932–1934, 93:89.
97 M. Jacobs 1932–1934, 97:44.
98 Harrington 1942, 22:1065b, 1171a.
99 Harrington 1942, 22:1281a.
100 Frachtenberg 1909; M. Jacobs 1932–1934, 91:9, 92:38, 98:28; Harrington 1942, 22:166.

CHAPTER 7

1 M. Jacobs 1932–1934, 91:5, 92:87, 100:42.
2 Harrington 1942, 23:225ab, 561.
3 Ward 1986:87.
4 M. Jacobs 1932–1934, 91:42; Drucker 1934; Maloney and Maloney 1933.
5 M. Jacobs 1932–1934, 91:42.
6 Harrington 1942, 23:452; Oregon State University Herbarium.
7 Haydon 1912.
8 Reg Pullen and Robert Kentta, personal communication, March 2, 2010.
9 Frachtenberg 1909; Harrington 1942, 24:586b, 964a; M. Jacobs 1932–1934, 92:133, 93:78.
10 Harrington 1942, 24:964a.
11 Harrington 1942, 22:785, 24:586a; M. Jacobs 1932–1934, 94:102.
12 Drucker 1934; M. Jacobs 1932–1934, 91:40–41.
13 E. Jacobs 1935, 120:73.
14 M. Jacobs 1932–1934, 91:40–41.
15 Drucker 1934; M. Jacobs 1932–1934, 91:31, 94:108; Sokolow 1965; E. Jacobs 1935, 120:73.
16 Harrington 1942, 24:742.
17 Drucker 1934; Barnett 1934; Maloney and Maloney 1933.
18 M. Jacobs 1932–1934, 95:57.
19 E. Jacobs 1935, 120:138; Robert Kentta and Bud Lane, personal communication, February 28, 2011.
20 Harrington 1942, 24:443a.
21 Harrington 1942, 23:704; M. Jacobs 1932–1934, 92:80.
22 Harrington 1942, 23:452.
23 Harrington 1942, 23:251a.
24 Frachtenberg 1920:89.
25 Earle and Reveal 2003:151.
26 Barnett 1934.
27 Ward 1986:25.
28 Drucker 1934; Pojar and MacKinnon 1994:109.
29 M. Jacobs 1932–1934, 93:93–94; Drucker 1934.
30 M. Jacobs 1932–1934, 91:92–93, 92:85, 93:56.
31 M. Jacobs 1932–1934, 93:56.
32 M. Jacobs 1932–1934, 93:58.
33 M. Jacobs 1932–1934, 100:48–50.
34 M. Jacobs 1932–1934, 92:93; Harrington 1942, 22:789; Drucker 1934.
35 M. Jacobs 1932–1934, 93:58.
36 Ward 1986:87.
37 M. Jacobs 1932–1934, 92:150.
38 Pojar and MacKinnon 1994.
39 Drucker 1934.
40 M. Jacobs 1932–1934, 91:38–39.

41 Harrington 1942, 22:577.
42 M. Jacobs 1932–1934, 97:50.
43 Harrington 1942, 21:754a, M. Jacobs 1932–1934, 91:39.
44 Harrington 1942, 21:754a.
45 Barnett 1934.
46 Harrington 1942, 22:796–97.
47 Harrington 1942, 21:844a; M. Jacobs 1932–1934, 92:13.
48 M. Jacobs 1932–1934, 100:64.
49 Moerman 1998:495–96.
50 Ward 1986:87.
51 Batdorf 1980:72.
52 Drucker 1934; Fluharty 2004; M. Jacobs 1932–1934, 93:93; Harrington 1942, 24:762a; St. Clair 1903.
53 Dorothy Kneaper, personal communication, November 2, 1999.
54 M. Jacobs 1932–1934, 100:104.
55 Gunther 1973:25.
56 M. Jacobs 1932–1934, 93:41, 91, 100:48–50.
57 Turner 1998:175.
58 M. Jacobs 1932–1934, 91:146.
59 Harrington 1942, 23:237ab.
60 Ibid.; M. Jacobs 1932–1934, 92:33.
61 M. Jacobs 1932–1934, 92:130; Harrington 1942, 22:748, 980b, 1023b.
62 Dorsey 1884.
63 Douglas 1914:227.
64 Dennis Rankin, personal communication, August 2, 2015.
65 Gunther 1973:25; Pojar and MacKinnon 1994:111.
66 Drucker 1934; Harrington 1942, 21:746, 753, 796, 22:799; Maloney and Maloney 1933; US Court of Claims 1931:87.
67 Haskin 1934:187, Turner 1995:91–92.
68 Grace Brainard, personal communication, November 12, 2000.
69 M. Jacobs 1932–1934, 91:144, 92:17.
70 M. Jacobs 1932–1934, 92:17, 32.
71 Grace Brainard, personal communication, November 12, 2000.
72 M. Jacobs 1932–1934, 93:52.
73 M. Jacobs 1932–1934, 100:64.
74 Grace Brainard, personal communication, November 12, 2000.
75 Harrington 1942, 24:246, 395ab.
76 Harrington 1942, 21:760; Ward 1986:86.
77 M. Jacobs 1932–1934, 94:118.
78 Harrington 1942, 23:922.
79 Harrington 1942, 22:725.
80 Harrington 1942, 24:589a.
81 Haydon 1912.
82 Drucker 1934.
83 Pojar and MacKinnon 1994:110.
84 Harrington 1942, 26:237.
85 Haskin 1934:23.
86 M. Jacobs 1932–1934, 93:59, 92–94, 100:26.
87 Harrington 1942, 22:775a; M. Jacobs 1932–1934, 93:59.
88 Harrington 1942, 25:722.
89 M. Jacobs 1932–1934, 98:85–89.
90 Wiedemann et al. 1999:78.
91 M. Jacobs 1932–1934, 91:38, 101:14; Drucker 1934; Harrington 1942, 22:783.
92 Harrington 1942, 22:783.
93 Ray Willard, personal communication, April 27, 2001.
94 Ward 1986:87.
95 Moore 1993:264.
96 M. Jacobs 1932–1934, 93:62.
97 M. Jacobs 1940:184–86.
98 M. Jacobs 1932–1934, 93:43.
99 Drucker 1934; M. Jacobs 1932–1934, 92:119.
100 Harrington 1942, 24:705b.
101 Ibid.
102 Lalande and Pullen 1999:262; Drucker 1934.
103 Lalande and Pullen 1999:262–63.
104 E. Jacobs 1935, 120:73.
105 Dickson 1946:46.
106 Harrington 26:226; Dorsey 1884.
107 M. Jacobs 1932–1934, 93:61, 100:48.
108 Harrington 1942, 22:584a, 794.
109 M. Jacobs 1932–1934, 92:37, 101:18; Harrington 1942, 22:790.
110 M. Jacobs, 1932–1934, 101:106; P. Whereat 2001.
111 Harrington 1942, 24:160b; M. Jacobs 1932–1934, 91:92–93.
112 M. Jacobs 1932–1934, 93:95.

113 M. Jacobs 1932–1934, 93:56.
114 M. Jacobs 1932–1934, 98:144.
115 M. Jacobs 1932–1934, 96:111.
116 M. Jacobs 1932–1934, 91:24, 41, 92:93; Harrington 1942, 22:231a; Barnett 1934; Drucker 1934.
117 Harrington 1942, 22:231a, 746.
118 Harrington 1942, 21:760; M. Jacobs 1932–1934, 92:31.
119 M. Jacobs 1932–1934, 92:31; Maloney and Maloney 1933.

CHAPTER 8

1 M. Jacobs 1932–1934, 91:37–38, 92:107; Maloney and Maloney 1933.
2 Harrington 1942, 22:784b.
3 M. Jacobs 1932–1934, 91:31.
4 M. Jacobs 1932–1934, 92:97.
5 Harrington 1942, 26:262.
6 Ward 1986:87.
7 Gunther 1973:13; Moerman 1998:424.
8 O'Neale 1995:27.
9 M. Jacobs 1932–1934, 92:79, 97, 102:107.
10 M. Jacobs 1932–1934, 91:24.
11 Ibid.
12 M. Jacobs 1932–1934, 101:22–24.
13 Harrington 1942, 24:965a.
14 M. Jacobs 1932–1934, 91:39.
15 Harrington 1942, 24:217.
16 M. Jacobs 1932–1934, 92:116; Harrington 1942, 22:1201b.
17 D. Hall and Phillips 2004:47.
18 M. Jacobs 1932–1934, 94:92; Harrington 1942, 22:1047a.
19 M. Jacobs 1932–1934, 92:49.
20 M. Jacobs 1932–1934, 97:44.
21 M. Jacobs 1932–1934, 92:88.
22 M. Jacobs 1932–1934, 91:9; Harrington 1942, 22:166.

CHAPTER 9

1 Harrington 1942, 22:770.
2 M. Jacobs 1932–1934, 99:39.
3 Gunther 1973:50.
4 Harrington 1942, 22:769a.
5 Harrington 1942, 24:737b.
6 Harrington 1942, 22:757.
7 Gunther 1973:50; Turner 1998:47.
8 M. Jacobs 1932–1934, 91:19; Maloney and Maloney 1933; Drucker 1934; Harrington 1942, 22:759.
9 Harrington 1942, 20:95a, 22:759.
10 Harrington 1942, 22:759.

CHAPTER 10

1 M. Jacobs 1932–1934, 91:41.
2 M. Jacobs 1932–1934, 94:102.
3 Gunther 1973:29.
4 Frachtenberg 1913:24; Golla 1965; Harrington 1942, 24:715a.
5 Douglas 1914:139.
6 Suckley and Cooper 1860, 55.
7 Brown 1868:380–1.
8 Drucker 1934.
9 Ibid.
10 M. Jacobs 1932–1934, 93:13.
11 Harrington 1942, 22:240.
12 Turner 1995:56.
13 M. Jacobs 1932–1934, 91:143.
14 M. Jacobs 1932–1934, 93:52–53.
15 Ibid.
16 Harrington 1942, 24:209.
17 M. Jacobs 1932–1934, 100:52.
18 M. Jacobs 1932–1934, 93:90.
19 Harrington 1942, 22:753a, 23:994, 24:748.
20 M. Jacobs 1932–1934, 91:65.
21 Harrington 1942, 24:723.

APPENDIX

1 M. Jacobs 1932–1934, 96:157–61.

Bibliography

Aikens, Melvin C., Thomas J. Connolly, and Dennis L. Jenkins. 2011. *Oregon Archaeology*. Corvallis: Oregon State University Press.

Anderson, M. Kat. 2005. *Tending the Wild: Native American Knowledge and the Management of California's Natural Resources*. Berkeley: University of California Press.

Annual Report of the Commissioner of Indian Affairs for the Year 1862. Washington, DC: Government Printing Office.

Annual Report of the Commissioner of Indian Affairs for the Year 1872. Washington, DC: Government Printing Office.

Barnett, Homer G. 1934. "Indian Tribes of the Oregon Coast." National Anthropological Archives, Smithsonian Institution, Washington, DC.

Batdorf, Carol. 1980. *The Feast Is Rich*. Bellingham, WA: Whatcom Museum of History and Art.

Beck, David R. M. 2009. *Seeking Recognition: The Termination and Restoration of the Coos, Lower Umpqua, and Siuslaw Indians, 1855–1984*. Lincoln: University of Nebraska Press.

Beckham, Stephen Dow. 1969. "Lonely Outpost: The Army's Fort Umpqua." *Oregon Historical Quarterly* 70 (3): 233–57.

———. 1977. *The Indians of Western Oregon: This Land Was Theirs*. Coos Bay, OR: Arago Books.

———. 1983. "Chronological Overview of the Historical Relationship of the Confederated Tribes of Coos, Lower Umpqua and Siuslaw Indians." Manuscript on file with the Confederated Tribes of Coos, Lower Umpqua, and Siuslaw Indians, Coos Bay, OR.

———. 1986. *Land of the Umpqua: A History of Douglas County, Oregon*. Roseburg, OR: Douglas County Commissioners.

———, ed. 2006. *Oregon Indians: Voices from Two Centuries*. Corvallis: Oregon State University Press.

Beckham, Stephen Dow, Kathryn Anne Toepel, and Rick Minor. 1984. *Native American Religious Practices and Uses in Western Oregon*. University of Oregon Anthropological Papers No. 31. Eugene: University of Oregon Department of Anthropology and Oregon State Museum of Anthropology.

Bensell, Royal A. 1959. *All Quiet on the Yamhill: The Civil War in Oregon.* Edited by Gunter Barth. Eugene: University of Oregon Books.

Blackburn, Thomas. 2005. "Some Additional Alexander W. Chase Materials." *Journal of California and Great Basin Anthropology* 25 (1): 39–54.

Boyd, Robert, ed. 1999. *Indians, Fire and the Land in the Pacific Northwest.* Corvallis: Oregon State University Press.

Brown, Robert. 1868. "On the Vegetable Products Used by the Northwest American Indians as Food and Medicine, in the Arts and in Superstitious Rites." *Transactions and Proceedings of the Botanical Society of Edinburgh* 9:378–96.

Christy, John A., James S. Kagan, and Alfred M.Wiedemann. 1998. *Plant Associations of the Oregon Dunes National Recreation Area.* US Department of Agriculture Forest Service Pacific Northwest Region Technical Paper, R6-NR-ECOL-TP-09-98.

DeLancey, Scott, and Victor Golla. 1997. "The Penutian Hypothesis: Retrospect and Prospect." *International Journal of Linguistics* 63 (1): 171–202.

Deur, Douglas, and Nancy J. Turner, eds. 2005. *Keeping It Living: Traditions of Plant Use and Cultivation on the Northwest Coast of North America.* Seattle: University of Washington Press.

Dickson, Evelyn M. 1946. "Food Plants of the Indians of Western Oregon." Master's thesis, Stanford University.

Dillon, Richard. 1975. *Siskiyou Trail: The Hudson's Bay Fur Company Route to California.* New York: McGraw-Hill.

Dodge, Orvil. 1898. *Pioneer History of Coos and Curry Counties, Or.: Heroic Deeds and Thrilling Adventures of the Early Settlers.* Salem, OR: Capital Printing.

Dorsey, James Owen. 1884. Dorsey Papers, Manuscript 4800. National Anthropological Archives, Smithsonian Institution, Washington, DC.

———. 1885a. "Lower Umpqua Vocabulary and Grammatical Notes." National Anthropological Archives, Smithsonian Institution, Washington, DC.

———. 1885b. "Siuslaw Vocabulary with Sketch Map Showing Villages." National Anthropological Archives, Smithsonian Institution, Washington, DC.

Douglas, David. 1914. *Journal Kept by David Douglas during His Travels in North America, 1823-1827: Together with a Particular Description of Thirty-Three Species of American Oaks.* London: William Wesley and Son.

Drew, E. P. 1858. Letter to Col. C. H. Mott, U. S. Commissioner, November 10, 1858. Oregon Superintendency of Indian Affairs Letterbooks, Roll 611, Microcopy 234, Frame 1132-41. National Archives, Washington, DC.

Drucker, Philip. 1934. "Ethnographic Field Notes." National Anthropological Archives, Smithsonian Institution, Washington DC.

DuBois, Cora. 2007. *The 1870 Ghost Dance*. Lincoln: University of Nebraska Press.

Earle, A. Scott, and James Reveal. 2003. *Lewis and Clark's Green World: The Expedition and Its Plants*. Helena, MT: Farcountry Press.

Fluharty, Suzanne M. 2003. "An Ethnobotanical Analysis of Basketry." Master's thesis, Oregon State University.

Frachtenberg, Leo J. 1909. "Coos Fieldnotes." Office of Anthropology Archives, Smithsonian Institution, Washington DC.

———. 1913. *Coos Texts*. Columbia University Contributions to Anthropology 1. New York.

———. 1914. *Lower Umpqua Texts and Notes on the Kusan Dialects*. Columbia University Contributions to Anthropology 4. New York.

———. 1920. *Alsea Texts and Myths*. Bureau of American Ethnology Bulletin 67. Washington DC: Smithsonian Institution.

Franklin, Jerry F., and C. T. Dyrness. 1973. *Natural Vegetation of Oregon and Washington*. Corvallis: Oregon State University Press.

Geary, E. R. 1860. Letter to Agt. Greenwood, October 1, 1860. Oregon Superintendency of Indian Affairs Letterbooks, Vol. G 10, M-2-8:223–25. National Archives, Washington, DC.

Gilkey, Helen M., and La Rea J. Dennis. 2001. *Handbook of Northwestern Plants*. Corvallis: Oregon State University Press.

Golla, Victor. 1965. "Hanis Word List." Manuscript on file with the Confederated Tribes of Coos, Lower Umpqua, and Siuslaw Indians, Coos Bay, OR.

Grant, Anthony P. 1994a. "Háanis tl'éeyis: The Hanis Coos Language; Some Common Words and Phrases and One Possible Way of Writing Them. With Notes on the History of the Language." Manuscript on file with the Confederated Tribes of Coos, Lower Umpqua, and Siuslaw Indians, Coos Bay, OR.

———. 1994b. "Námłiitnxam wá'as (Our Language): The Siuslaw and Kuitsh Language." Manuscript on file with the Confederated Tribes of Coos, Lower Umpqua, and Siuslaw Indians, Coos Bay, OR.

Guard, B. Jennifer. 1995. *Wetland Plants of Oregon and Washington*. Renton, WA: Lone Pine.

Gunther, Erna. 1973. *Ethnobotany of Western Washington: The Knowledge and Use of Indigenous Plants by Native Americans*. Seattle: University of Washington Press.

Hall, Debra, and Patricia Phillips. 2004. *Ethnobotany of the Coos, Lower Umpqua, and Siuslaw: Plants Used for Tools, Food, Medicine and Clothing as Remembered by Our Elders and Ancestors*. Coos Bay, OR: Confederated Tribes of Coos, Lower Umpqua, and Siuslaw Indians.

Hall, Roberta L. 1984. *The Coquille Indians: Yesterday, Today and Tomorrow*. Lake Oswego, OR: Smith, Smith and Smith.

Harrington, John P. 1942. "Alsea, Siuslaw, Coos, Southwest Oregon Athapaskan: Vocabularies, Linguistic Notes, Ethnographic and Historical Notes." John Peabody Harrington Papers, Alaska/Northwest Coast. National Anthropological Archives, Smithsonian Institution, Washington, DC.

Haskin, Leslie. 1934. *Wild Flowers of the Pacific Coast.* Portland, OR: Binfords and Mort.

Haydon, Walton. 1912. "List of Plants of Coos Bay." Typescript in possession of Stephen Dow Beckham, Lake Oswego, OR.

Hines, Gustavus. 1852. *Oregon: Its History, Condition and Prospects: Containing a Description of the Geography, Climate and Productions with Personal Adventures among the Indians*. Buffalo, NY: George H. Derby.

Hymes, Dell. 1966. "Some Points of Siuslaw Phonology." *International Journal of American Linguistics* 32:328–42.

———. 2003. *Now I Know Only So Far: Essays in Ethnopoetics.* Lincoln: University of Nebraska Press.

Jacobs, Elizabeth. 1935. "Upper Coquille Notes." Notebook 119, Jacobs Collection. University of Washington Libraries, Seattle.

———. 2003. *The Nehalem Tillamook: An Ethnography*. Edited by William Seaburg. Corvallis: Oregon State University Press.

Jacobs, Melville. 1932–1934. "Coos Ethnologic Notes," Notebooks 91–99, 101, Jacobs Collection. University of Washington Archives, Seattle.

———. 1939. *Coos Narrative and Ethnologic Texts.* Seattle: University of Washington Press.

———. 1940. *Coos Myth Texts.* Seattle: University of Washington Press.

Jensen, Edward C., and Charles R. Ross. 2005. *Trees to Know in Oregon.* Corvallis: Oregon State University Press.

Kasner, Leone Letson. 1976. *Siletz: Survival for an Artifact.* Dallas: Itemizer-Observer.

Kozloff, Eugene N. 1983. *Seashore Life of the Northern Pacific Coast: An Illustrated Guide to Northern California, Oregon, Washington, and British Columbia*. Seattle: University of Washington Press.

———. 2005. *Plants of Western Oregon, Washington and British Columbia.* Portland, OR: Timber Press.

Kramer, Stephenie. 2000. "Camas Bulbs, the Kalapuya, and Gender: Exploring Evidence of Plant Food Intensification in the Willamette Valley of Oregon." Master's thesis, University of Oregon.

Krumm, Bob, and James Krumm. 1998. *The Pacific Northwest Berry Book: A Complete Guide to Finding, Harvesting and Preparing Wild Berries and Fruits in the Pacific Northwest.* Helena, MT: Falcon.

Lalande, Jeff, and Reg Pullen. 1999. "Burning for a 'Fine and Beautiful Open Country': Native Uses of Fire in Southwestern Oregon." In *Indians, Fire and the Land in the Pacific Northwest*, edited by Robert Boyd. Corvallis: Oregon State University Press.

Losey, Robert J., Nancy Stenholm, Patty Whereat-Phillips, and Helen Vallianatos. 2003. "Exploring the Use of Red Elderberry (*Sambucus racemosa* ssp. *pubens*) Fruit on the Southern Northwest Coast of North America." *Journal of Archaeological Science* 30:695–707.

Maloney, Alice, and Joe Maloney. 1933. "Coos Ethnographic Notes from Joe and Alice B. Maloney." Melville Jacobs Papers. Special Collections, University of Washington Libraries, Seattle.

Mithun, Marianne. 1999. *The Languages of Native North America.* Cambridge: Cambridge University Press.

Moerman, Daniel E. 1998. *Native American Ethnobotany*. Portland, OR: Timber Press.

Moore, Michael. 1993. *Medicinal Plants of the Pacific West.* Santa Fe, NM: Red Crane Books.

Moss, Madonna L. 2005. Tlingit Horticulture: An Indigenous or Introduced Development? In *Keeping it Living: Traditions of Plant Use and Cultivation on the Northwest Coast of North America*, edited by Douglas Deur and Nancy J. Turner, 274–95. Seattle: University of Washington Press.

Moss, Madonna L., and Jon Erlandson. 1995. *An Evaluation, Survey, and Dating Program for Archaeological Sites on State Lands of the Northern Oregon Coast, with Reports on Archaeological Surveys of South Slough (Coos Bay) and of Intertidal Fishing Sites*. Report of the Department of Anthropology, University of Oregon, to the State Historic Preservation Office, Salem.

Nelson, Nancy. 2000. "The Umpqua Eden Site: The People, Their Smoking Pipes and Tobacco Cultivation." Master's thesis, Oregon State University.

Nisbet, Jack. 2012. *David Douglas: A Naturalist at Work*. Seattle: Sasquatch Books.

Odeneal to Case, May 30, 1873, in US Office of Indian Affairs, *Letters Received by the Office of Indian Affairs, 1824–1880*. National Archives Microcopy 234, Roll 618:589–94.

O'Neale, Lila M. 1995. *Yurok-Karuk Basket Weavers*. Berkeley: Phoebe A. Hearst Museum, University of California.

Oregon State University Herbarium. http://www.oregonflora.org/atlas.php.

Pojar, Jim, and Andy MacKinnon. 1994. *Plants of the Pacific Northwest Coast: Washington, Oregon, British Columbia, and Alaska*. Redmond, WA: Lone Pine.

Schultz, Stewart T. 1990. *The Northwest Coast: A Natural History*. Portland, OR: Timber Press.

Schwartz, E. A. 1991. "Sick Hearts: Indian Removal on the Oregon Coast, 1875–1881." *Oregon Historical Quarterly* 92:228–64.

Seaburg, William. 1994. "Collecting Culture: The Practice of Ideology of Salvage Ethnography in Western Oregon, 1877–1942." PhD diss., University of Washington.

Sokolow, Jane. 1965. "Hanis Word List." Manuscript on file with the Confederated Tribes of Coos, Lower Umpqua, and Siuslaw Indians, Coos Bay, OR.

St. Clair, Harry Hull. 1903. "Coos Field Notes." Office of Anthropology Archives, Smithsonian Institution, Washington, DC.

Suckley, George, and J. G. Cooper. 1860. *The Natural History of Washington Territory and Oregon with Much Relating to Minnesota, Nebraska, Kansas, Utah and California between the Thirty-Sixth and Forty-Ninth Parallels of Latitude*. New York: Baillière Brothers.

Todt, Donn L. 1997. "Cross-Cultural Folk Classifications of Ethnobotanically Important Geophytes in Southern Oregon and Northern California." *Journal of California and Great Basin Anthropology* 19 (2): 250–59.

Turner, Nancy J. 1995. *Food Plants of Coastal First Peoples*. Vancouver: University of British Columbia Press.

———. 1997. *Food Plants of Interior First Peoples*. Vancouver: University of British Columbia Press.

———. 1998. *Plant Technology of First Peoples in British Columbia*. Vancouver: University of British Columbia Press.

———. 2014. *Ancient Pathways, Ancestral Knowledge: Ethnobotany and Ecological Wisdom of Indigenous Peoples of Northwestern North America*. Montreal: McGill-Queen's University Press.

Tveskov, Mark Axel. 2000. "The Coos and Coquille: A Northwest Coast Historical Anthropology." PhD diss., University of Oregon.

US Court of Claims. 1931. Testimony in *Coos et al. v. United States*, Docket K-345. National Archives, Washington, DC.

Ward, Beverly H. 1986. *White Moccasins*. Myrtle Point, OR: Myrtle Point Printing.

Whereat, Donald. 2011. *Our Culture and History: The Confederated Tribes of the Coos, Lower Umpqua and Siuslaw Indians*. With Patty Whereat Phillips, Melody Caldera, Ron Thomas, Reg Pullen, and Stephen Dow Beckham. Newport, OR: Donald Whereat.

Whereat, Patricia. 2001. "Tobacco: Indigenous Practices and Uses on the Southern Oregon Coast." In *Changing Landscapes: Proceedings of the Fourth Annual Coquille Cultural Preservation Conference, 2000.* Edited by Jason Younker, Mark A. Tveskov, and David G. Lewis. North Bend, OR: Coquille Indian Tribe.

Wiedemann, Alfred M., La Rea J. Dennis, and Frank H. Smith. 1999. *Plants of the Oregon Coastal Dunes.* Corvallis: Oregon State University Press.

Winter, Joseph, ed. 2000. *Tobacco Use by Native North Americans: Sacred Smoke and Silent Killer.* Norman: University of Oklahoma Press.

Youst, Lionel. 1997. *She's Tricky like Coyote: Annie Miner Peterson, an Oregon Coast Indian Woman.* Norman: University of Oklahoma Press.

Index